The
Mathematical Nature
of the Living World

THE POWER OF INTEGRATION

The
Mathematical Nature
of the Living World

THE POWER OF INTEGRATION

Gilbert Chauvet

Ecole Pratique des Hautes Études & CHU Angers, France

World Scientific

NEW JERSEY · LONDON · SINGAPORE · BEIJING · SHANGHAI · HONG KONG · TAIPEI · CHENNAI

Published by

World Scientific Publishing Co. Pte. Ltd.

5 Toh Tuck Link, Singapore 596224

USA office: 27 Warren Street, Suite 401-402, Hackensack, NJ 07601

UK office: 57 Shelton Street, Covent Garden, London WC2H 9HE

British Library Cataloguing-in-Publication Data
A catalogue record for this book is available from the British Library.

THE MATHEMATICAL NATURE OF THE LIVING WORLD
The Power of Integration

ISBN-13 978-981-238-819-3
ISBN-10 981-238-819-2

Typeset by Stallion Press
Email: enquiries@stallionpress.com

CONTENTS

Prolog

UNDERSTANDING THE FUNCTIONING OF LIVING ORGANISMS

An Introduction to Theoretical Integrative Physiology

Prolog

UNDERSTANDING THE FUNCTIONING OF LIVING ORGANISMS

An Introduction to Theoretical Integrative Physiology

"Even in the imperfect, Nature has usually sketched in the outline of perfection, so that it is only by degrees that we attain the light."
Marcello Malpighi (1628–1694)

Today, giant telescopes explore the frontiers of the universe, powerful synchrotrons crunch the tiniest particles of matter, and sophisticated molecular biology techniques probe the secrets of the human genome. Nevertheless, in spite of all the extraordinary achievements to the credit of modern science, the great riddle of life, and its relationship to nonliving matter, remains unsolved.

The mysteries of the origin and the evolution of life add to those of the physical universe. Biological evolution includes not only the apparent transformation of one species into another, but also the progression of an individual organism from conception to death. Thus, a better understanding of evolutionary mechanisms may be expected to lead to new insights into the phenomena of life. Scientists have long striven to understand the organization of the living being. The rapid development of physics in the twentieth century allowed biologists to borrow some interesting ideas from thermodynamics and cybernetics, such as the principle of biological order stemming from a pre-existing physical order (Schrödinger, 1945), or arising from thermodynamic fluctuations (Prigogine, Nicolis & Babloyantz, 1972), or even being initiated by random noise (Atlan, 1972). Although these ideas are difficult to apply directly to the reality of functional biology, they have helped to elaborate some of the general principles that govern living organisms.

Our main objective here is to develop a theory of functional biology incorporating the notion of the unity of living organisms. This approach is meant to establish the fundamental principles underlying the functioning of living organisms, and to try to explain how life arose in nonliving matter. The theory is based on the novel concept of *functional interactions* that serve as elementary building blocks to construct the entire edifice. The heuristic method used calls for highly abstract, mathematical concepts. Just as physics uses mathematics to provide a general view of the nonliving world, biology will have to rely on mathematical formalism to obtain an integrated vision of living organisms. Furthermore, we shall see how the use of mathematical tools helps us to deduce the evolutionary possibilities of a given species from available physiological knowledge. However, let the reader have no fear — scarcely any mathematics will appear in this book.[1]

The determination of the characteristics of life in nonliving matter requires that biology be separated from physics, while recognizing that biological phenomena are based on physical substrates. On one hand, the science of physics appears to be a magnificent monument constructed on the bedrock of unity; indeed, the beauty of the physical world is reflected, for instance, in the unique mathematical operator that simultaneously describes all the energy levels of matter. On the other hand, the science of biology is perceived as an enormous, confused mass of loosely related information with hardly any unity, with the exception of the fields of immunology and molecular genetics that are just beginning to establish certain basic principles.

---

[1]The formal description of physiological mechanisms on the basis of models, integrated in a synthetic "bottom-up" approach, was first published in a three-volume work: "*Traité de Physiologie Théorique*", by G. Chauvet, Masson, Paris (1987–1990). A revised and updated version is available in an English translation by K. Malkani, under the title: *Theoretical Systems in Biology: Hierarchical and Functional Integration*, Pergamon, Oxford (1996).

The fundamental lack of unity in our knowledge of biology becomes particularly evident under conditions of extreme physiological distress, as in the case of a patient receiving intensive medical care. The critical situation calls for an approach integrating all the molecular data available, together with the vast amount of information concerning the different physiological functions of the body. What is then really needed is a general unification of the biological disciplines that, for historical reasons, have developed along distinctly separate lines.

We may observe that our very claim to existence depends on the fact that the elements of matter we are composed of, together with our thinking processes, constitute a *stable* singularity in the universe we live in. Of course, this stability must have a cause, but can anyone say what it is? Putting aside the usual stereotypes, can we distinguish living organisms from material objects? Can the phenomena of life be reduced to mere physicochemical reactions?

Let us recall that it took humankind thousands of years to develop the concept of the movement of bodies in space and to define the notion of physical force. In fact, this represents a highly abstract construction based on the observation of a very common phenomenon. There is an enormous gap between the perception of a moving object and the idea of the force acting on it, similar to that between the material description of the object and its mental representation. This is analogous to the difference between the material description of protein biosynthesis and the abstract description of the underlying functional processes in terms of a set of hierarchical interactions. As we shall see, the main advantage of this abstract approach is that it can be generalized to all the physiological phenomena observed in the living organism.

The phenomena of respiration as viewed by Aristotle and Claude Bernard

While the physicochemical structure of living matter and a wide range of physiological phenomena in living organisms are now well known, we still cannot describe the *global nature* of living matter or give a satisfactory definition of *life*. To do so, we would need to describe the integrated functioning of living organisms in terms of basic principles expressed in mathematical terms. Thus, the essential problem posed is that of the integration of the different structural levels — genes, organelles, cells, tissues and organs — culminating in a coordinated set of physiological functions.

To put the idea of the integration of physiological functions into a historical perspective, let us consider brief extracts from the writings of two of the great contributors to physiology: Aristotle (384–322 BC) and Claude Bernard (1813–1878).

Here is what Aristotle wrote about respiration in animals:

"Let us now speak of [...] the causes of respiration for this is what allows certain animals to keep alive while others die. But why do animals that have a lung receive and breathe air, particularly those that have a lung filled with blood? The cause of respiration is that the lung is spongy and filled with vessels; this is the part, among those called the viscera, which contains the most blood. However, all the animals that have a lung filled with blood need to be cooled rapidly since the vital fire exercises considerable influence, penetrating the body of the animal because of the quantity of blood and heat present. Air satisfies this requirement since, because of its subtle nature, it can move quickly to all parts of the animal; water would have the opposite effect. Now why, among the animals that breathe, is it principally those with a lung filled with blood that do so? The answer to this question is evident considering that what is warmer needs more cooling and that the breath moves easily towards the principle of heat situated in the heart." (Aristotle, *Parva Naturalia,* c.330 BC).

In contrast, here is Claude Bernard's point of view on respiration in general, expressed twenty-five centuries later:

"Air is necessary to the life of all living organisms, plants or animals; air thus exists in the internal organic atmosphere. The three gases of the external air: oxygen, nitrogen and carbon dioxide, are in the dissolved state in the organic liquids in which the histological elements breathe directly, like fish in water. The cessation of life due to a lack of the gases, particularly of oxygen, is what is called death by asphyxiation. In living organisms, there is a constant exchange between the gases of the internal medium and the gases of the external medium; but as we know, plants and animals are not alike with respect to the changes they produce in the environment. [...] I then studied the action produced by carbon dioxide on blood by artificial poisoning. [...] These experiments, repeated under identical conditions, taught me that in such situations there was a simple, volume for volume exchange between carbon dioxide and oxygen in the blood. However, the carbon dioxide, while displacing the oxygen expelled from the blood, remained fixed to the red corpuscles such that neither the oxygen nor any of the other gases could displace it. Thus, the death of the organism was consequent upon the death of the red corpuscles. In other words, death occurred because of the cessation of the exercise of the physiological property, essential to life, of the red corpuscles." (Claude Bernard, *Introduction à l'étude de la médecine expérimentale*, 1865).

The logic of Aristotle's explanation is based on the *examination of facts*, whereas that of Claude Bernard relies on *experimental observation*. Aristotle was a convinced finalist and, although his conclusion may seem somewhat quaint today, we must admire the audacious criticism he brought to bear on such great predecessors as Democritus, Empedocles and Plato, before putting forward his own views based on known facts. What is even more important than his insightful explanation is the extraordinary philosophical and methodological impact of his thought. In contrast, Claude Bernard, who was an ardent determinist, had at his disposal all the knowledge of the physical sciences accumulated over the centuries. Moreover, the considerable technological advances of his time allowed him to plan and carry out sophisticated experiments. Here is what he wrote about the experimental method he applied to the problem of blood poisoning:

"At first I knew absolutely nothing about the mechanism of the phenomenon of blood poisoning by carbon dioxide. I experimented to see, or rather to observe, what happened. The first observation concerned the specific modification of the color of the blood. I interpreted this observation and then made a hypothesis that was proved false by an experiment. However, this experiment provided another observation, which I used as the starting point for constructing a new hypothesis for the mechanism of the removal of oxygen from blood. By constructing successive hypotheses, each time in keeping with the observed facts, I was finally able to demonstrate that carbon dioxide replaces oxygen in the red corpuscle, after combining with the substance of the cell."

By analyzing the characteristics of the experimental method that enabled him to lay the foundations of physiology, Claude Bernard proved that he was not only a pioneering experimental scientist but also a sound theoretician. Like Descartes, he claimed to rely only on the authority of observed facts and experimental demonstrations, independently of any personal bias. He would first generalize a hypothesis in the form of a theory that would then serve as the starting point for experimental research. The essence of Claude Bernard's approach is best summed up in his own statement:

"When we come upon a fact that is in contradiction with an established theory, we must accept the fact and abandon the theory, even if the theory is generally upheld and applied by the greatest of scientists."

What is the present situation in physiological research? A century and a half after Claude Bernard, such a vast number and diversity of incontestable facts have been collected that several new disciplines have sprung up, such as embryology, molecular biology and neuroscience, to name but a few. Many partial biological theories have been elaborated in Claude Bernard's sense, i.e. with the generalization of experimentally tested hypotheses within the framework of a given biological discipline. However, is the number of partial theories now sufficient to attempt an explanation, or at least the beginning of an explanation, of the global functioning of the living organism?

While it is relatively simple to explain the functioning of certain parts of an organism, such as the gene, the neuron or the

heart, it is much more difficult to approach the integrated functioning of the organism by means of qualitative theories or
hypotheses. Nevertheless, from the beginning of the 20th century, several interesting attempts have been made. Thus, from
D'Arcy Thompson (1942) to Thom (1972) and Murray (1989),
several scientists have investigated the theoretical aspects of
morphogenesis, i.e. the development of the forms of living
organisms. The groundbreaking work by Grödins, Buell & Bart
(1967) on the physiology of respiration was followed, thanks
to the development of powerful computers, by research on the
major physiological systems (see, for example, Yates (1982)).
Among these scientists, Rashevsky (1961), whose work
deserves to be better known, should be singled out for his fundamental contribution to relational biology, which constitutes
an original theoretical approach to the functioning of living
organisms.

Ideally, what is needed today is a representation in which
formal logical reasoning could be applied to a multitude of
variables, such as the pressure of body fluids, the concentration of hormones, the frequency of the emission of neuronal
potentials, and so on. Thus, Claude Bernard's ideas of general
physiology will have to be developed in the form of a new
discipline that we may call integrative physiology. Currently,
theoretical physiology gives a quantitative interpretation of
observed facts within a framework limited to a subsystem of the
organism, usually represented by only a small number of variables. In contrast, the theory of integrative physiology may be
expected to provide a qualitative interpretation for the entire
system, composed of its several subsystems. In other words,
the new theory should be able to describe phenomena involving
very large numbers of variables. The investigation of individual
organs could then be extended to the study of the organism as
a whole.

Although Claude Bernard clearly recognized that mathematical laws subtend biological reality, the enormous difficulties
then inherent to the formal representation of living organisms
led him to take a prudent stand. He suggested it would be

more useful to continue the search for new facts, rather than put those that were already known in the form of mathematical equations. Thus, he wrote:

> "If we wish to understand the laws of biology, we should not only observe and record the phenomena of life but also determine the numerical values of the intensity of the relationships between them. This application of mathematics to natural phenomena is the aim of all science since the expression of the laws governing the phenomena must be essentially mathematical. Therefore, the data subjected to calculations should be obtained from facts that have been sufficiently analyzed so that the phenomena entering the equations are fully known. However, in the case of most biological phenomena, I believe that attempts of this kind are premature, precisely because these phenomena are so complex that, in addition to the few conditions that are known, we may be certain that there are many still absolutely unknown. Thus, the most useful approach in physiology and medicine would lie in searching for new facts rather putting into equations those facts already known. I definitely do not condemn the application of mathematics to biological phenomena since it is only through mathematics that biological science will eventually be established. However, I am convinced that it is impossible to formulate a general equation for the time being. In fact, the qualitative study of biological phenomena must necessarily precede the quantitative study."

Today, the number of biological facts recorded is enormous and, in an increasing number of cases, the phenomena observed just cannot be understood without a formal approach. For example, the phenomenon of memory, one of the main functions of the nervous system, would be incomprehensible without a sound mathematical explanation, such as that given by Hopfield (1982).

Is the time now ripe for integrative physiology?

With hindsight, we see how accurately Claude Bernard had prophesized the fantastic development of experimental biology. Today, one of the greatest challenges of biology lies in the integration of the multiple facts observed. The usual approach relies on the use of biomathematical methods for the construction of specific biological models. A well-known example is that of the Hodgkin–Huxley model (Hodgkin & Huxley, 1952) for the nervous system. The novelty of the approach we propose here lies in the integration of observed facts into a formal theoretical framework that would allow the explanation of biological phenomena by means of a small number of general principles.

In short, we believe that theoretical biology will do for biological science what theoretical physics has done for physical science, i.e. provide a formal approach that, by integrating models specific to the different levels of the living organism, will enable us to understand its global nature. Of course, this does not mean simply superimposing physical models onto biological phenomena since, as we shall see, biology cannot be reduced to physics. The main drawback to formalized integration in biology is that we would have to use specific mathematical tools within an abstract conceptual framework in which each variable, whether observable or not, possesses two distinct aspects: the physical aspect, because of its structure; and the biological aspect, because of its function.

In spite of the difficulties involved, this approach may be expected to lead to the discovery of novel biological properties and enable us to answer some fundamental questions. For instance, how do we define a physiological function in general, operational terms? Again, how do we distinguish between a living organism and a physical system? These questions are far from trivial since they concern some of the essential aspects of life.

A novel approach to the functioning of living organisms

The theory we propose concerns not only the organization, but also the functioning of a living organism, i.e. it incorporates the topology as well as the geometry of the body. It is based on the existence of *functional interactions* between the biological structures of the organism, and on that of the *hierarchical organization* of the body, from the molecular level to that of the whole organism. As we shall see, functional interactions have two fundamental properties, called *non-symmetry* and *non-locality*, which play an essential role in the development of living organisms.

How do we represent such functional interactions with their non-symmetrical and non-local properties? One method consists of using graph theory to represent the functional hierarchy induced by the resultants of the functional interactions. Another method, complementary to the graph theory-based method, consists of applying field theory to represent the spatiotemporal changes of the oriented, non-symmetrical action, i.e. the transfer of the product, after its transformation, from the source to the sink. In mathematical terms, a field is a quantity that varies at each point of space. We may define each functional interaction by means of an operator that, at a given instant and at a given point in physical space (the source), propagates the variable quantity towards another point (the sink). This abstract description of the dynamics of action implicitly takes into account the anatomy of the living organism, while explicitly describing the propagation of the product, after its transformation, from one structural unit to another.

An important question then arises: Why do functional interactions appear at all in living organisms? Before reaching adulthood, all organisms go through a developmental phase during which their structural and functional organizations are modified according to their genetic program. Why is this program executed as actually observed? Putting aside the ideas of finalism — the philosophical doctrine that claims there must be an underlying reason for each existing structure — we may wonder why two structures might associate with each other when, on one hand, this leads to an increase in the complexity of the system and, on the other, each of the two structures could *potentially* exist independently. As we shall see, unlike physical systems in general, the stability of a biological system — at least under certain conditions — results from its complexity, and the stabilization is due to the association of structures. In biological terms, the maintenance of homeostasis during the development of a living organism produces associations between its structures, thereby increasing the stability of its physiological dynamics.

Another equally crucial question concerns the origin of the functional hierarchy of living organisms. We may ask why, at a given moment during the development of a biological organism, should there be a preferential extension of a particular hierarchical organization? A similar problem, encountered in physical systems, is resolved by the *principle of least action.* This principle states that the motion of a conservative dynamical system between two points occurs such that the action has a minimum value along a specific path, compared to all the other possible paths. The true reason for the obligatory "choice" of this path lies in the *geometry* of the space in which the movement occurs. As we shall see in the case of a biological system, the choice is linked to the stability of the functional organization, i.e. the *topology* of the system, in other words, it depends on the stability of the relationships between sources and sinks. Why does one structural unit become a source, and another a sink? This fundamental biological problem raises many questions awaiting satisfactory answers.

During the development of an organism by cell differentiation, we observe one particular functional organization rather than some other that might seem equally possible. This implies the existence of a potential of organization, in other words, there must be a reservoir of potentialities available for the organization of the system, and there must be an underlying cause for the final organization. What determines the privileged value of the potential of this organization? It could be, for instance, the number of cells that exist at a given point of development, as is evident, or again the number of sources and sinks of the functional organization, as is somewhat less evident.

Furthermore, during the development of an organism, the number of structural units varies because of cell division. How are the structural units then reorganized? How does specialization appear? Is there a criterion for the evolution of biological systems? The hierarchical organization must develop such that the physiological function remains invariant during the successive transformations of the functional organization. Thus, as in the case of physical systems, there must exist a functional order, mathematical in essence, that governs the evolution of biological systems.

The approach we propose to use in addressing these questions is based on a level of abstraction not commonly used in biology. In a sense, it stems from a thought experiment since, paradoxically, we have to consider the non-observable to understand the observable. This is similar to certain applications of the principle of least action in physics that oblige us to consider certain impossible situations, for example, that of light rays *not* being propagated in the shortest possible time between two points. This kind of reasoning, which aims at determining optimal principles, has proved extremely fecund in physics. Furthermore, the very nature of the questions that arise in biology invites us to use methods that are analogous to those used in physics.

However, in the field of biology, particularly in the case of human biology, we are immediately confronted with further, highly mysterious questions with profound ramifications: Why are we what we actually are? Why are we in the particular state of development in which we find ourselves, and not in some other? Does a potential for biological organization really exist? If it does, then under other circumstances we might have been quite different from what we actually are. In physical terms, this would imply that biological evolution follows an optimal path, maintaining the essential properties of living species.

The examples given in the following chapters, selected from various biological fields such as embryology, neurobiology and cardiovascular physiology, as well as from the fundamental phenomena of birth, life and death, will illustrate the necessity of laying down the foundation of a new theoretical biology, and explain the meaning we attach to this term. From the fecundated egg to the embryo, from the embryo to the adult organism, which finally falls victim to the irreversible aging process, the destiny of the individual appears to be ineluctable and *determined*, at least in the biological sense. *Thus, if the notion of determinism can be applied to the phenomena of evolution, we have to try to elucidate the underlying principles governing the evolution of the individual living organism, in particular, and that of life, in general.*

The living organism, in spite of all its differences with non-living matter, is part of the physical universe and must naturally be subject to physical laws. It is therefore difficult to believe the biological world to be devoid of the unity that the wonderful harmony of the laws of the physical universe entices us to seek — the elusive unifying theory. Just as the non-living world affords a mental representation of a certain physical reality, it should be possible to construct a mental representation of the reality of the living world. This construction is bound to be a gigantic task, perhaps even impossible to realize fully. In any case, it will call for a formidably high level of mathematical abstraction.

The history of science teaches us that our knowledge of the universe comes from multiple discoveries, experimental as well as theoretical, followed by the synthesis of the partial results within the framework of a formal theory. True, it is difficult to be sure that the time has now come to attempt such a synthesis in biological sciences. It might justifiably be felt that it is illusory to seek to integrate biological phenomena as diverse as development and growth, learning and memory, aging and irreversibility, into a single conceptual framework.

Thus, the reader should not expect too much. We certainly cannot claim to offer a complete physiological explanation of living systems, nor deliver the magic key to the vast, unexplored domain of the global functioning of biological organisms. Rather more modestly, we hope to contribute a few elements towards the mathematical formulation of physiological functions, and thence deduce, from a small number of concepts, a theoretical representation. The validity of the concepts used will then have to be tested against the consequences of the theory. Of course, as is often the case with theories, whatever proposed here will have to be revised according to future results obtained.

Prolog

One of the main objectives of a theory of integrative biology would be to compare biological and physical systems. We shall therefore introduce certain physical concepts such as field theory and the notion of entropy. However, it should be made clear that the genesis of the biological theory, far from being derived by analogy with any physical theory, is the result of a deliberate effort to integrate individual physiological functions into a consistent, formalized framework.

In his biography of the great precursor of modern scientific thinking, Galileo Galilei (1564–1642), the French historian of science, Geymonat (1982), has this to say:

"I believe that, all things considered, the distinction between the truth deduced by hypothesis and the truth based on fact, a distinction already present in the other writings of Galileo, finally provides us with the true key to the profound meaning of his whole work. [...] If this is true, I might well be asked: How do you explain that Galileo decided to dedicate such a large part of his research to the study of motion? How do you explain that he struggled so hard to demonstrate through deduction what he might have been satisfied to prove by experiment alone? Before answering these questions, we must observe that, in making the distinction between the mathematical demonstration of a 'property of motion' and the experimental verification of this property, we do not deny that the former may, in a way, sustain the latter. In itself, the mathematical demonstration may not be able to establish the truth of the property, but it will be able to throw light on the logical links between one property and another, thereby enabling us to grasp the general principles involved in the conclusions brought up by the experiment."

Thus, four centuries ago, Galileo had already felt the necessity of combining the results of the theoretical and the experimental investigations of physical phenomena to elucidate the general underlying principles. We can hardly afford to adopt a different attitude today, especially with the modern computing techniques now available to test new theoretical ideas in the complex field of biology. Of course, it would be unrealistic to expect to obtain the full interpretation of biological phenomena in the immediate future. Let us not forget that it took physics several centuries of hard thinking to explain the law of attraction between two bodies. In the field of biology, we will have to create specific mathematical tools, construct physiological models, design numerical simulations, and then elaborate a general theory applicable to the multiple systems of the living organism. This Herculean task will require the collaboration of much more than the mere handful of theoretical biologists that are now working along these lines. The identification of the functional interactions of the human body surely deserves the setting-up of ambitious research programs, at least on the same scale as those dedicated to exploring the human genome or the frontiers of the human brain, to mention only the more recent international projects in human biology.

I

PHYSICAL AND BIOLOGICAL INTERACTIONS

Does Physical Force have a Biological Counterpart?

I

PHYSICAL AND BIOLOGICAL INTERACTIONS

Does Physical Force have a Biological Counterpart?

Some scientists believe that the *human intelligibility of nature is fundamentally mathematical.* In other words, the apparent harmony of our universe is subtended by mathematics, the understanding of which alone can lead to a synthesis of the diversity of the phenomena observed. In fact, the history of physics has been marked by powerful syntheses such as Maxwell's theory of electromagnetism, Einstein's theory of relativity and, more recently, Hawking's theory of superstrings, which unifies the preceding theories. All these ideas emerged after long periods of scientific gestation that produced the mathematical formulation required to explain the phenomena of light through electromagnetic principles (Maxwell, 1872), gravitation through geometrical concepts (Einstein, 1923), and the evolution of the universe through gravitational laws and the quantum theory (Hawking, 1988). Rather unfortunately for most of us, the further we advance in science, the more abstract our image of the physical universe becomes, i.e. the mental representation we construct of it, try as we might to keep it as simple as possible (Espagnat, 1980).

Can physical and biological interactions be compared?

A very commonly observed physical interaction in nature is that of the force exerted by one body upon another. What exactly does the abstract notion of force correspond to? Physics deals with relatively simple systems that can be broken down into smaller systems, allowing an easier approach to the problem. In comparison, biological systems, as we shall see, are far more complex. One of our main objectives here is to distinguish the living from the nonliving. An important difference between physics and biology lies in the number of levels used for the description of natural phenomena. In a physical system, we need consider no more than two or three levels of description. For example, the properties of an electric current flowing in a conductor are explained by the transport of free particles, i.e. the electrons, in the conductor, for instance a copper wire. If we admit that the organization of particles of matter constitutes a structure, then we see that this explanation is based on only two levels of description: the macroscopic level (the electrical activity in the copper wire) and the microscopic level (the flow of electrons). In other words: on one hand, we have the set of copper atoms and, on the other, copper atoms with their peripheral electrons.

In a given physical system, the interplay between the sets of influences acting on the particles regulates the equilibrium of the system. These influences, called *forces*, stabilize the physical structure. The science of physics deals essentially with the forces acting between the particles, and their consequences, in solids and fluids, or even *within* elementary particles. Today, it appears evident that when a force acts on a particle and causes it to move, the system is in dynamic equilibrium. However, this only came to be understood when Newton first expressed the idea in a general, inductive form in his *Philosophiae naturalis principia mathematica* (1687). In his formulation, based on geometry, he enunciated three laws. The first corresponds to *Galileo's principle of inertia,*[2] which states that a body in motion

---

[2]Galileo Galilei, *Discouri* (Dialogues and mathematical demonstrations concerning the two new sciences of mechanics and local motion), (Leyden 1638). An interesting review of Galileo's life and work will be found in Geymonat's *Galilée* (1992).

continues moving at a uniform speed in a straight line, unless acted upon by an external force. The second law determines the variation of the velocity of the body under the action of an external force, the force being defined as the product of the mass of the body, which characterizes its matter, and the variation of its velocity, or acceleration. The third law states that the force F acting between two bodies varies directly as the product of the masses of the bodies, m and m', and inversely as the square of the shortest distance r between them. In mathematical terms, we have the equation: $F = mm'/r^2$. With his third law, Newton created the theory of gravitation. This was certainly a stroke of genius since the third law not only lays down the rule under which one body influences another but also gives a qualitative description of the influence by introducing the abstract concept of the *force* involved in the physical interaction.

Newton's third law of motion expresses an idea that is at once simple and extraordinary. The concept of the force is valid not only for bodies in motion on the Earth but also for heavenly bodies. Moreover, the idea of the gravitational force describes the fact that one mass can act instantaneously on another at a distance, apparently with no transfer of any kind. It is this mysterious, non-local character of the law of gravitation, almost bordering on the occult, that caused Leibnitz, the rationalist, to oppose Newton, who insisted that far from trying to answer the *why* he was only addressing the *how* of the phenomena of gravitation. According to Newton's third law, the force of gravitational origin acting between two masses, called the gravitational interaction, satisfies the principle by which the action exercised by one particle A on another particle B is equal to that exercised by B on A. As readily seen in the equation above: the body of mass m being at x and the body of mass m' being at x', changing $m(x)$ to $m'(x')$ does not alter the intensity of the force since it is the square of the distance $r = |x - x'|$ that is involved. This is the *principle of action and reaction* contained in the law of gravitation. Another, more "operational" way of expressing this principle is to say that the momentum of a body, equal to the sum of the momentums of all its constitutive bodies, is conserved during its motion. The momentum is equal to the product of the mass of the body and its velocity. This principle of conservation explains the mechanism of propulsion of a rocket, for example, in which case the mass m_A of the combustion gases ejected at a velocity v_A is equal to the mass m_B of the rocket propelled at a velocity v_B in the opposite direction, since $m_A v_A + m_B v_B = 0$. Since the mass of the gases is much smaller than that of the rocket, the velocity of the gas ejected will have to be proportionally high to propel the rocket at the required velocity.

The mathematical expression of Newton's laws of gravitation is rather simple, although the application of the laws to a system comprising several interacting bodies can be a formidable task. However, the fact that these laws can be formulated mathematically gives them an enormous advantage over other laws that cannot — at least for the time being — be so expressed. Thus, in human physiology, most of our knowledge can only be stated in qualitative terms, such as: "low temperatures lead to an increase in diuresis", or "emotional shocks trigger the secretion of adrenalin into the blood". The mathematical formulation of a law, in addition to providing quantitative prediction, allows the practical deduction of all the consequences. In the case of Newton's laws, this ensemble of consequences constitutes the classic science of mechanics. For example, the movement of a satellite around the Earth brings into play at least two bodies, the Earth and the satellite; but we also have to consider the action of the other planets of the solar system, together with their own satellites, as well as that of all the other celestial bodies in the universe. Thus, the application of even relatively simple physical laws to a system comprising more than two bodies soon becomes inextricable.

In his celebrated essay on the three-body problem, Poincaré (1890) proved the impossibility of resolving the equations of movement for three bodies, for example, the Sun, Jupiter and Saturn, while the problem could be readily solved for two bodies. Finally, the Kolmogoroff-Arnold-Moser theorem, developed between 1954 and 1967, showed that the stability of a three-body system could be determined only by imposing perturbations on the system. Thus, when a Hamiltonian system, i.e. a frictionless system, is subjected to a slight perturbation, although several initial conditions lead to a quasi-periodic, stable trajectory, a few can lead to a chaotic trajectory. This notwithstanding, one of the foundations of the organization of the physical structure of matter, composed of subatomic particles, atoms and molecules, is built on the abstract concept of *force*, and on the general, almost tautological principles of the conservation of the momentum and the electric charge of bodies.

Do biological systems involve elementary interactions playing a functional role analogous to that of a force in a physical structure? Among the many questions raised by the growth and development of a living organism, we may wonder how a cell "recognizes" the state of its development at a given moment, and how it "decides" to react to a particular situation. It is likely that, at the gene level, sequential and parallel sets of instructions, similar to those of computer programs, determine the precise instant at which a specific protein is synthesized. In turn, this protein will trigger a set of enzymatic reactions leading to other sets of biochemical reactions. Such sets, incorporating mechanisms of synthesis and their corresponding regulation, may be very complex indeed. Whether we consider the processes of embryonic development, or those involved in memory and learning, or in the regulation of physiological functions in higher organisms, the common point is the existence of molecules or signals emitted by one cell, called the *source*, acting at a distance on another cell, called the *sink*. This is illustrated, for instance, by the propagation of an electric potential along a nerve, the action of a hormone operating at a distance from its site of synthesis after being carried in the bloodstream, and the change produced in the shape of a molecule after it binds to another molecule. Table I.1 gives some examples of such interactions observed in human physiology.

These examples enable us to grasp a concept that is fundamental to the understanding of biological phenomena: the concept of the *functional interaction*. In simple terms, this implies that *a product emanating from one entity acts on another at a distance*. Thus, a signal emitted by a given cell acts on another cell, which in turn emits a new signal after transforming the signal received. This defines the functional interaction between two cells since the action operates from one cell to the other.

Table I.1: Some examples of functional interactions in human physiology.

Physiological function: propagated signal	Source	Sink
Nervous activity: *potential*	A neuron	Another neuron
Thyroid metabolism: *thyroxin*	A thyroid cell	All cells
Allosteric activity: *change of shape*	A molecular subunit	Another molecular subunit
Gene regulation: *repressor*	A regulator gene	A structural gene
Pulmonary ventilation: *partial pressure of O_2*	A pulmonary region	Another pulmonary region

It is called "functional" because the function of the first cell is to *act* on the second by means of the signal it emits, causing the second cell to *produce* a new signal at a distance.

The functional interaction greatly resembles a mathematical function in that it consists of the application of one set on another, transforming an element of one set into an element of the other set. For this reason, we shall use the term *functional interaction* indifferently to describe the *action* of one biological structure on another as well as the *product* of this action, just as in mathematics we identify the function $f\colon x \to f(x)$ with its value $f(x)$. In fact, the functional interaction is the most elementary physiological function, a "building-block" with which the entire functional organization may be constructed. In this sense, and only by analogy with the construction of physical systems, we may say that the role of the functional interaction in biology is similar to that played by force in physics.

However, it is important to note the nature of the difference between the biological functional interaction, on one hand, and physical interactions, gravitational or electrical, on the other, as described below.

The gravitational interaction is *symmetrical*, in the sense of action and reaction, since it is expressed by the force: $F = \kappa \, mm'/r^2$, which, by linking two objects of masses m and

m', creates a physical structure, in other words, a dynamic assembly of matter. Similarly, the electrical interaction is also *symmetrical*, since it is expressed by the same type of equation: $F = \varepsilon qq'/r^2$, where the electric charges on the bodies replace the masses in the gravitational equation. Without going further into the fascinating question of the role played by symmetry in physics, a subject that has already received considerable attention (Wigner, 1970; Coleman, 1985), we would like to stress the profound influence of simple principles on the formulation of a problem. The property of symmetry must necessarily appear somewhere in the formulation even though, in the case of a highly complex system, it may not be as evident as in the formula: $F = mm'/r^2$.

Consider, for example, the energy of vibration and rotation of a molecular system due to the permanent vibration and rotation of the atoms of a molecule about their equilibrium position. This energy must remain invariant when the system undergoes symmetrical operations. Thus, in the case of a molecule composed of two vibrating atoms, such that the geometry remains unchanged when the atoms are interchanged symmetrically, say about a median plane, the energy must remain constant since the molecule remains identical to itself. The formulation of the property of the invariance of energy must therefore include the notion of symmetry. The property of invariance is reflected by the appearance of energy levels[3] that correspond exactly to the symmetry of the molecule. In mathematical terms, the symmetry groups[4] to which the molecule belongs define its energy levels. In practice, infrared or Raman spectroscopy is used to measure

---

[3]These energy levels can be calculated by quantum mechanics. They are given by the eigenvalues of the Hamiltonian operator representing the energy of the vibration and rotation of the molecular system. The operator, expressed in the form of a matrix, is factorized according to the representations of the group of symmetry. On simplification, zero-values appear in the matrix at specific points corresponding to the symmetry of the molecule.

[4]In mathematics, a *group* consists of a set associated with an operation, satisfying certain elementary operations. This mathematical structure, due to Evariste Galois (1811–1832), has proved very useful for the formalization of problems not only in physics but also in other fields, such as that of linguistics.

the energy levels of a molecule and establish its group of symmetry. This actually constitutes a kind of molecular fingerprint whereby the molecule is identified (Wilson, Decius & Cross, 1955). This method has been successfully applied even to large biological molecules, such as adenosine triphosphate (Chauvet, 1974). Methods based on symmetry are commonly used in physics since the property of symmetry is closely linked to the structural organization of matter, which is by essence of a geometrical nature. Thus, the laws of conservation of matter are associated with the properties of symmetry of the system. A law of conservation of matter is considered to represent a fundamental property of the system to such an extent that, in physical theories, the knowledge of the laws of conservation is considered sufficient to explain the system.

In contrast to physical interactions, the functional biological interaction is *non-symmetrical*. This property of non-symmetry should therefore be incorporated in the formulation of biological interactions. The biological functional interaction operates from one structural unit on another situated at a distance, i.e. from a source to a sink, although not *directly* from source to sink. This interaction is unidirectional since the molecule or signal, emitted at a given level of organization, will have no direct retroaction from sink to source. The consequences of the non-symmetry of the functional interaction on the development of biological systems and on the dynamics of biological processes in general are discussed in Chapters V and VI. As we shall see, these dynamics are governed by equations that include the property of non-symmetry.

The other important property of the functional biological interaction is that of *non-locality*, since the product emitted is transported at finite speed from source to sink. The property of non-locality is as fundamental to living organisms as that of locality associated with physical systems.[5] We shall see in Chapters V

---

[5]Putting it simply, we may say that a local phenomenon observed here and now may be deduced from what happened here just a little while ago, whereas a non-local phenomenon observed here and now has to be deduced from what happened far away from here and perhaps quite some time ago.

and VI, how the notion of non-locality, first introduced in physics by Brillouin (1952), is related to the representation of biological phenomena and, more precisely, how it results from the continuity underlying hierarchical systems (Chauvet, 1993a).

The concept of functional biological interactions raises a number of interesting questions. For instance, how do two functional interactions differ? The answer is that the difference is defined by the "nature" of the products of the interactions, i.e. the ions, molecules, hormones or potentials emitted, just as the difference between physical interactions is defined by the nature of the forces in play, be they gravitational, electrical or nuclear. However, in the case of two biological interactions, the properties of non-symmetry and non-locality have similar consequences and, in particular, lead to the structuring of the biological system in the form of hierarchical levels of organization of the physiological functions. This is analogous to the way the property of symmetry leads to the structuring of a molecular system with its different energy levels, though without any hierarchical organization. Thus, the properties of non-symmetry and non-locality lead to a novel situation in theoretical biology, obliging us to formulate physiological functions incorporating the idea of a hierarchical organization that must be reflected by the mathematical solution of the dynamic biological system.

Another question that may be asked is what happens when a given biological unit acts as a source for certain products and as a sink for others? This problem can be tackled mathematically and, as we shall see, the solution leads to other fundamental questions. We may also ask what happens when a source turns into a sink, or vice versa. In fact, this leads to the appearance of new properties related to the functional reorganization of the organism. Here also the dynamics of processes associated with this particular topology lead to specific properties. We shall address these questions in the following chapters and try to determine the elements that are common to physiological phenomena. We may then be able to deduce a fundamental criterion of biological evolution. To sum up, the functional biological interaction, an abstract mathematical entity representing, for instance, an electrical potential, a molecular concentration, a saturation level or some other bio-logical parameter, involves three elements: the *source*, the *sink* and the *transformation* within the sink, and possesses at least the two properties of *non-symmetry* and *non-locality*.

However, why should this particular representation of biological phenomena in terms of functional interactions be considered more appropriate than some other? The simple answer is that the smallest number of concepts common to the largest number of biological phenomena may be expected to lead us to the largest number of properties interpretable by means of these concepts. In other words, the smallest number of very general principles should allow us to interpret the largest number of phenomena observed and, in particular, those that are still unexplained. This, of course, is the objective of the theory we are trying to construct. The goal is far from being attained, and we do not even know if it will ever be reached. It is true there are other possible representations of biological phenomena, such as those based on the concept of compartments. Considering the importance of this concept, we shall discuss it below in connection with that of the functional interaction proposed here.

Some consequences of the concept of functional interactions

The somewhat abstract considerations above allow us to deduce the structure of a general biological law involving functional interactions, at least in the case of *formal* biological systems defined on the basis of a small number of properties. Such for-mal systems correspond to an idealization of real biological sys-tems. Physical laws governing gravitational interactions reveal the property of symmetry through the mathematical expression of the product of the masses and the relationship with the square of the distance between them. The symmetry of electri-cal interactions is expressed in a similar manner. Again, we find symmetry in the expression for entropy given by statistical mechanics in the description of molecular collisions. In contrast to physical laws, biological laws will have to explicitly include the property of non-symmetry between the sources and sinks involved in biological functional interactions. Then, if we admit that functional interactions constitute the basis of biological phenomena, we may go on to describe some of the consequences.

The first consequence concerns the mathematical properties of functional interactions. Sources, sinks and their interactions can be conveniently represented by a *mathematical graph*[6] (Figure I.1). Two points on a sheet of paper correspond to the source and the sink of a functional interaction, and the interaction itself is indicated by an oriented arc joining the two points, with an arrow on the arc to distinguish the source from the sink. If we generalize this procedure to any number of sources and sinks, we obtain a set of summits connected by oriented arcs, possessing mathematical properties. The summits are called "structural units" and the number of summits corresponds to the "degree of organization" of the structure. As we shall see, this kind of graph can be used to represent the functional organization of a biological organism. In fact, it constitutes the *topology* of the biological system. In other words, it corresponds to the *functional organization of a formal biological system* (O-FBS).

Let us now briefly recall some notions concerning *topology*. Founded by Listing in 1847, this branch of mathematics was

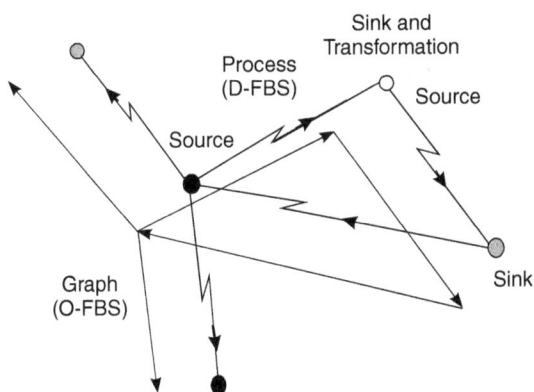

Figure I.1: A superposition of the graphs representing the topology of the organization of a formal biological system (O-FBS) and its dynamics (D-FBS).

---

[6]In the sense of the mathematical graph theory.

greatly developed by Poincaré (1890).[7] Topology studies the properties of space in terms of the notion of *neighborhood*, and consequently that of *continuity*. For example, objects that differ from the point of view of Euclidean geometry, such as the triangle, the circle or the square, are considered identical from the topological point of view. In contrast, a disk with a hole in it and a disk without a hole will be considered topologically distinct since the one cannot be transformed into the other by means of any continuous deformation. When we consider the position of points relative to each other without taking into account the distances between them, we obtain a topological graph. In other words, topology is what is left of geometry when the notion of the metric is removed. The connectivity between sources and sinks in a functional system is thus of a topological nature, but the interactions between them depend not only on the topology but also on the geometry of the system.

Mathematical graphs have often been used in fields as varied as economics, physics and biology to describe the complex relationships between the elements. Although these graphs, consisting of points and arcs, are apparently identical, they differ because of the properties imposed by the problem to be solved. Let us consider some examples.

The well-known puzzle of the Koenigsberg bridges proved to be one of the starting points of the mathematical theory of graphs (Bergé, 1983). The problem was to find a path in the town crossing each of its seven bridges only once. This corresponds to class of practical situations, e.g. the traveling salesman's problem, in which the traveler's route has to be optimized. Such problems are readily solved by combinatorial analysis and graph theory.

In a very different context, complex chemical systems may be analyzed by using *network thermodynamics* (Oster, Perelson and Katchalsky, 1973), the theory of which is derived from the generalization of the thermodynamics of irreversible processes

---

[7]See the Collected Works of Poincaré (1980) for a full account of his contribution to topology.

(Glansdorff and Prigogine, 1970). The method introduces the use of *binding graphs* to calculate energy exchanges between the different elements of a given chemical system. In fact, the underlying principle is the unification of various scientific disciplines from a thermodynamic point of view. Thus, it has been applied to the transport of mass within biological tissues, as well as in the modeling of highly complex physicochemical phenomena in biology. In the case of binding graphs, the arcs represent the paths along which the energy circulates.

Another example is that of the *interaction graphs*, in which the arcs represent the circulation of matter. The formalism applied to these graphs is the same as that of the transformation systems (Delattre, 1971a), which allowed the extension of compartmental analysis to the phenomena of radiation. As we shall see further on, compartmental analysis is also widely used in biology (Jacquez, 1985). Mathematical graphs have many applications, for instance, in electrical networks for the transport of energy, in Kirchoff's laws for the conservation of current at the junctions in an electric circuit, in telecommunications networks for the transport of information, in rail and road networks for the transport of goods, and so on. In short, mathematical graphs are very convenient for the representation of the displacement of matter or energy between the different reservoirs of a system, as well as for the analysis of the resulting global dynamics (Roy, 1970).

The second consequence of the existence of functional interactions concerns the formal expression of the dynamics of the process corresponding to the spatiotemporal fate of the product that, after being emitted by a source, will act at a distance on a sink. For example, a hormone is transported at a rate depending on the blood flow, a neurohormone migrates along the axon of a hypothalamic neuron for several days before reaching its target, an action potential is propagated at a velocity of eighteen meters per second along the larger nerve fibers, and so on. Since the delay with which the signal from the source reaches the sink depends on the localization of the source and the sink, the geometry of the biological system influences the dynamics of the process and is therefore implicitly included in this representation. The functional interaction being determined by its three elements, i.e. the source, the sink and the transformation of a product in the sink, the propagation of the signal and its action at a distance represent the dynamics of the process, describing *how* the interaction takes place (Figure I.1). Thus, the determination of the *dynamics of the formal biological system* (D-FBS), i.e. the representation of the action, in space and in time, of one structural unit of a system on another is clearly complementary to that of the *functional organization of the formal biological system* (O-FBS), i.e. its topology.

The *field theory* used in physics offers a convenient method for describing the propagation of the signal and its action at a distance. According to this theory, each point in physical space can be associated with a certain numerical value that is characteristic of the process analyzed. Consider, for example, the value at each point of an electric potential being propagated point by point over the membrane of a neuron. Such values may also be time-dependent. Thus, the action at a distance may be described by the spatiotemporal variation of a field, its value at a given point-sink being the result of the action of a *field operator* at the point-source at which the signal was emitted. Such values are represented by a *field variable* corresponding to the functional interaction, or more precisely, to the mathematical value of the functional interaction. This formulation in terms of the field theory may seem rather abstract but, as we shall see in Chapter VI, it offers considerable advantages in the study of biological systems.

To sum up, *the functional interaction is characterized by two complementary aspects, one being the triplet composed of the source, the sink and the arc between them, and the other being the spatiotemporal field describing the action of the source on the sink. The combinatorial analysis of the triplets gives the topology of the biological system, i.e. its functional organization, whereas the variation of the field determines the manner in which the products are exchanged in time and in space, thus defining the dynamics of the biological process.*

The use of graph theory to describe the O-FBS, combined with the use of the field theory to describe the D-FBS, are the two conceptual frameworks within which the execution of a set of functional interactions may be described. The combination of the two dynamic systems,[8] one corresponding to the O-FBS and the other to the D-FBS, describes the spatiotemporal operation of the biological system. Here, we have an important difference between physical and biological systems. Unlike physical systems, *in biological systems the topology, represented by the functional connections between the structural units, and the geometry, represented by the density of the structural units at each point in physical space, may both vary with time.*

For example, during biological development, the cellular density at a given point in the organism varies according to the rate of cell division. At the same time, new tissues are formed and new functional connections are established. Consequently, the dynamics of the new products exchanged undergoes a modification, bringing about a functional reorganization involving the topology as well as the geometry of the biological system. Again, when a biological organism suffers aggression from an antigen, the antibody-producing cells of the immune system first multiply rapidly to face the danger, and then decrease in number when normal conditions are restored. It is true that in physical systems we may observe analogous phenomena, for example during *phase transitions,* but these concern only the structural organization of matter without affecting its functional organization.

The conceptual framework proposed above allows the formalization of a certain type of problem. For example, a perturbation of the topology of a biological system will affect its evolution by provoking a reorganization of the functional interactions between its structures (O-FBS) while reestablishing the dynamics of the processes involved (D-FBS). The dynamic stability, which expresses the return of the system to

---

[8]In mathematical terms, a dynamic system is a set of differential equations describing a process, or the operation of a phenomenon, between two successive instants: t and $t + dt$. In this case, we refer to the local dynamics, since the state of the dynamic system depends only on its immediately preceding state at a neighboring point. Of course, the biological system should be distinguished from the corresponding dynamic system.

its starting point after having been perturbed, also corresponds to the stability of its constituent systems, i.e. the O-FBS and the D-FBS. These two aspects of a formal biological system, each acting on the other, are inextricably linked. At a higher conceptual level, even a phenomenon as complex as the evolution of a biological species may be considered as the resultant of a coupling between the topology, the geometry and the dynamics of the associated processes.

How do biological principles differ from physical principles?

Biology differs from physics not only because of the nature of the entities studied by these sciences, but also because of the nature of the principles underlying the explanation of the phenomena observed. Whereas only a few principles of quantum physics and relativity are required for the construction of mathematical theories governing a wide range of physical phenomena, there are no such unifying principles in biology. Of course, biology has its own laws, for example those of the evolution of the species by selection and mutation, the doctrines of molecular biology, the principles of physiology, and so on. However, for the most part these laws are not mathematical and even when they are, as in the case of certain optimum principles in cardiovascular or respiratory mechanics, they do not lead to any truly *integrated* understanding of the functioning of a biological organism or its evolution.

Considering the differences between physics and biology, few biologists would willingly rely on physical models alone to explain biological phenomena. However, putting aside the mystical ideas of vitalism and admitting that all forms of life — from viruses to mammals — are but matter, it is not unreasonable to hold that living organisms must obey general laws of organization and functioning that differ from physical laws. We believe that the difference between living organisms and non-living matter may be accounted for by a theory constructed from general principles in a manner that has always proved satisfactory in the domain of physics. Moreover, such a theory would allow the simulation of biological phenomena by means of elementary mechanisms, thereby providing better insight into the processes involved. In fact, living organisms, no matter how complex, display the property of *stability* during their functioning as well as their development. In other words, the physiological state of the organisms is maintained in the neighborhood of a stationary state — either that of equilibrium or of one far from equilibrium. This kind of stability strongly suggests the existence of a few general principles in functional biology.

What exactly is a principle? Several laws, such as those of gravity, electromagnetism, hydraulics, nuclear interactions, and so on, govern physical phenomena. These laws share common points that may be stated in the form of great general principles. The principles of conservation of matter, momentum, energy, electric charge, the principles of symmetry and relativity, and the principle of least action, are a few of the examples in physics. These principles are so general and valid in so many different situations that they may actually appear tautological. However, as Poincaré (1854–1912) indicated in his essay on *Science and Hypothesis* (Poincaré, 1968), this is far from being the case:

> "What I have just said also sheds light upon the role of general principles such as the principle of least action or that of the conservation of energy. These principles are of the greatest value since they have been determined by seeking all that was common in the enunciation of several physical laws. In a way, they represent the quintessence of innumerable observations. However, their very generality leads to the consequence that they can no longer be verified. Since we cannot give a general definition of energy, the principle of the conservation of energy simply means that something remains constant. Then, no matter what new notions future experiments might bring, we may already be certain that there will be something that will remain constant — something that we may call energy. Does this mean that the principle has no meaning and that it disappears into a tautology? Of course, it does not; it means that the different things we call energy are in fact linked by a true relationship. However, if this principle has a meaning, it may be false. Perhaps we do not have any right to extend the applications of

the principle indefinitely even though we may be sure it will be verifiable in the strict sense of the term. But how shall we know when the principle has been extended to its legitimate limit? Quite simply, when it ceases to be useful, i.e. as soon as it fails to provide a correct prediction for some newly observed phenomena. We may then be sure that the relationship claimed is no longer real, since otherwise the principle would still be fecund. Such an observation, without directly contradicting a further extension of the principle, will have nevertheless pronounced its death sentence."

In *The Value of Science*, Poincaré went on to add:

"These principles result from the generalization of experimental observations, but from this very generality they seem to borrow a remarkable degree of certitude. In fact, the more general the principles the more often they can be tested, and the multiplication of the results in all their various forms — even the most unexpected — finally leaves no room for doubt."

Thus, in physics we have the *principle of least action*, which appears to be extraordinary, probably because of its apparently anthropomorphic nature. The principle states that a particle will choose a path between two points such that the *action* has a minimum value with reference to all the other possible paths between these points. This would suggest that the particle actually makes a mathematical decision based on the comparison of numbers corresponding to the paths along which it has never really traveled. This principle, which is valid not only in mechanics but also in optics and electromagnetism, may thus appear highly mysterious. But the mystery is quickly solved if we admit that the path of the particle depends on the geometry of the space it moves in, and that the particle moves along a *geodesic line*, i.e. the shortest distance between two points according to the nature of the space. For example, this would evidently be a straight line in a plane, or an arc of the great circle on the surface of a sphere.

We have mentioned that the main advantage of a principle in physics is its generality. Thus, Newton's laws of motion can be replaced by the unique, strictly equivalent postulate of the *variational principle* that is generally applicable in physical mechanics. This new formalism is elegantly simple and yet powerful enough to be used for complex systems. Furthermore, it can be effectively extended to disciplines other than that of Newtonian mechanics, such as quantum mechanics and classic field theory. As we shall often refer to this concept in the chapters that follow, it may be useful to give a brief description here.

Formalisms in analytical mechanics

In mechanics, a system that is *free*, i.e. one with no external force acting on it, and *conservative*, i.e. one in which the total energy is constant, obeys two fundamental laws. The first is the law of conservation of momentum since, on one hand, the action and the reaction being equal and opposite cancel the sum of the internal forces taken two by two, and on the other, the system being free, the resultant of the external forces is null. The second is the law of conservation of energy during the movement, since the work done by the forces is independent of the paths taken by the material points of the system, depending only on the starting and finishing points of the movement. In other words, in a conservative system the total mechanical energy E, i.e. the sum of the kinetic energy T and the potential energy V, is a constant: $E = T + V =$ constant. According to Leech (1961):

> "The practical importance of the concept of energy lies in the fact that all the mechanical properties of a complex system can be summed up by putting into analytical form a certain number of scalar functions designated by the term energy. Analytical mechanics is based on the generalization of this idea."

The principles of conservation express the fact that a certain quantity calculated according to a given rule will always give the same result whatever the natural conditions. In the physical universe, some quantities (such as energy, electrical charge and the number of particles) are conserved, whereas other quantities (such as frictional energy and heat) are not, but the final balance must satisfy a law of variation with time. *Conservative systems* are of great importance from a theoretical point of view since they are relatively simple to formulate in terms of general principles. Thus, it can be shown that the forces acting on a conservative system depend upon the potential function in a very special way. These forces may be considered as being derived from a potential,[9] such that $F = -\partial V/\partial x$. In contrast to conservative systems, almost all real physical systems are *dissipative systems*. In such systems, there is a spontaneous loss of energy, as in the case of an oscillating pendulum, which will eventually stop moving because of the frictional forces at work. This is why dissipative systems require a specific formulation.

Furthermore, mechanical systems may be subject to other complications such as the limitation of movement imposed by conditions of a geometrical nature, materialized by *liaisons*. In the case of the pendulum, the weight must move over a constant distance at each oscillation. Such liaisons introduce additional unknown forces into the equations of movement. Although the notion of a liaison is an abstract, mathematical concept, it facilitates the resolution of problems in mechanics. As Leech (1961) put it: the notion of a liaison is but the simplified image of a more complex, real phenomenon. To deal with these complications, a new postulate going beyond Newton's laws of motion was laid down: the *principle of virtual work*. This principle states that *the total virtual work of all the forces acting on a system of material points is null, whatever be the virtual, infinitesimal movements of these points, starting from an equilibrium state of the system*. In other words, the work done by the forces of liaison acting in a system is positive or null for all virtual displacements compatible with the liaisons. Here again we have a principle that is not easy to accept immediately. Although it deals with virtual displacements, i.e. displacements that are not real, the principle actually allows us to describe reality and has proved indispensable in the formulation of several problems in analytical mechanics.

---

[9]For example, water flowing downhill will follow the path corresponding to the "greatest slope", implying the operation of an optimum principle.

The two main principles of mechanics, due to the work of *d'Alembert* (1717–1783) and *Hamilton* (1805–1865), are applicable far beyond the field of Newtonian mechanics. According to d'Alembert's principle, *the total virtual work of the effective forces, i.e. the applied forces and the forces of inertia, acting on a system of material points is null, whatever be the infinitesimal, virtual movements, reversible and compatible with the liaisons between these points.* From d'Alembert's principle, which may be considered as the most condensed and general statement of the laws of mechanics, we can deduce not only Newton's laws of motion and Hamilton's principle but also the law of the conservation of energy when the virtual movement coincides with a real movement.

The formalism due to Lagrange (1736–1813) is strictly equivalent to that of Newton. The equations of motion are deduced from the function: $L = T - V$, called the Lagrange function or the Lagrangian of the system. This function depends on the parameters of position q, which in turn depend on the coordinates of positions x, y and z. A system composed of N particles will therefore be governed by $3N$ equations, called the Lagrangian equations. This formalism can also be applied to a non-conservative system by replacing V with a specific function M to take into account the dissipative nature of the system. An example of this is the analysis of the motion of a charged particle in an electromagnetic field.

The Hamiltonian formalism adds coordinates p, called *conjugated momentums*, to the parameters of position q. These supplementary coordinates, defined by the relationship: $p = \partial L/\partial q$, correspond to quantities of movement when the system is conservative. Whereas q and $\dot{q}$ are dependent, since $\dot{q}$ is the derivative of q with respect to time, the variables p and q can be considered independent. Thus, the $3N$ equations of the Lagrangian formalism are replaced by the $6N$ equations of the Hamiltonian formalism. The trajectory of a system can then be represented by that of a point having $6N$ coordinates in a space with $6N$ dimensions, called the *phase space*. The N trajectories in physical space are thereby replaced by a single trajectory in phase space. One of the major advantages of the Hamiltonian formulation is that it provides a perfect framework for quantum mechanics as well as statistical mechanics. As in the case of the Lagrangian formalism, the Hamiltonian formalism introduces the Hamilton function: $H = \sum_{pq} - L$, also called the *Hamiltonian*, which generally represents the total energy of the system. For a conservative system, it can be shown that: $H = T + V = E$, if the transformation of the Cartesian coordinates into the coordinates of position, i.e. $x \equiv x(q)$, does not depend explicitly on time. The equations of motion written using the Hamiltonian formalism, called the canonic equations[9], may also be used for a dissipative system. However, in this case the Hamiltonian H will no longer correspond to the total energy of the system, which is certainly one disadvantage of the formalism.

The mysterious principle of least action

The background sketched in above allows us to present the *variational principles* that we propose to apply to the important problem of the evolution of biological systems. Variational principles, commonly used in physics, are noteworthy for the elegance of their expression and the ease with which they can be used to solve physical problems. Let us consider the origin of variational principles. Figure I.2 shows that, given a function: $y = f(x)$, movement may take place from one point $A(x_1, y_1)$ to another $B(x_2, y_2)$ along different paths, such as (C) or (C′). Each of these paths may be assigned a certain value represented by

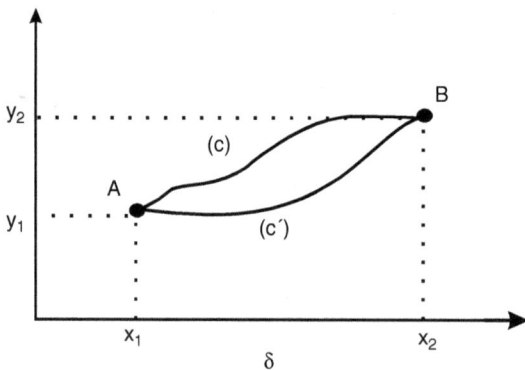

Figure I.2: The principle of least action for a null variation with time. The extremum curve passes through A and B.

the function I, and a path may be selected such that the function I be a stationary function, i.e. a function for which the variation, or the first-order derivative, is null. This particular path, called the *extremum curve*, corresponds to the *extremum function*. The variation is denoted by the symbol δ so that the inertia of the extremal function I is given by the expression: $\delta I = 0$. This function, which represents the global effect between two points A and B, in fact integrates a succession of local effects, say F, between two points: x et $x + dx$, that are infinitesimally close together.

The extremum curve represents the shortest path under conditions of minimum stationariness. In most problems, the extremum is required to be a minimum, but in some cases the extremum desired may correspond to the maximum. In mechanics, a problem enunciated in these terms leads to an equation similar to that of Lagrange,[10] but in which the Lagrangian L is replaced by the function F described above. Then the function I, in this case called the *Hamiltonian action*, noted S, represents the integration of the quantities: $L = T - V$, between the two instants corresponding to the beginning and the end of the movement. *Hamilton's principle* establishes that the variation δ of the action is null, or $\delta S = 0$. Thus, Hamilton's principle, which is a consequence of Newton's laws of motion, may be considered as a fundamental principle from which the Lagrange equations and the whole field of mechanics may be derived by means of a concise and elegant formulation. Applied to Newtonian dynamics, Hamilton's principle corresponds to a principle of least action.

Let us now consider another type of *principle of least action*, which has several interesting applications. This is obtained from a more general variation, noted Δ. The variation Δq is null at the limits q_1 and q_2, whereas the variation Δt is not null (Figure I.3).[11] Then, if the system is conservative, the principle of least action leads to a new function W, called the *Maupertuis function*, of

---

[10]The Lagrange equation is written: $\dfrac{\partial L}{\partial q} - \dfrac{d}{dt}\left(\dfrac{\partial L}{\partial \dot{q}}\right) = 0$, with $\delta \int_{t_1}^{t_2}(q,\dot{q},t) = 0$.

[11]The relationship between the two variations is given by: $\Delta = \delta + \Delta t \dfrac{d}{dt}$

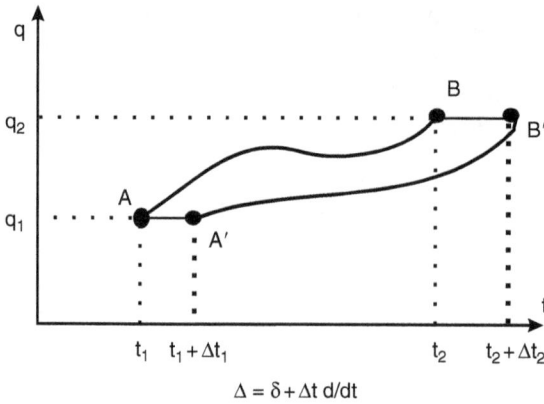

$$\Delta = \delta + \Delta t\, d/dt$$

Figure I.3: Here there is a variation of time as well as a variation of position: the variation includes the displacements *AA'* and *BB'*.

which the variation Δ is null.[12] This is a more general form of the principle of least action since it takes into account the variation of time as well as the variation of the parameters of position *q*.

The principles indicated in Table I.2 are of great importance in dynamics since they express a very subtle, fundamental physical property. As Feynmann (1969) said in one of his informal lectures on physics:

"Problem: Find the true path. Where is it? One way, of course, is to calculate the action of millions and millions of paths and look at which one is lowest. When you find the lowest one, that's the true path.

That's a possible way. But we can do it better than that. When we have a quantity that has a minimum, for instance, in an ordinary function like the temperature — one of the properties of the minimum is that if we go away from the minimum in the *first* order, the deviation of the function from its minimum value is only *second* order. At any place else on the curve, if we move a small distance the value of the function changes also in the first order. But at a minimum, a tiny motion away makes, in the first approximation, no difference.

That is what we are going to use to calculate the true path. If we have the true path, a curve that differs only a little bit

---

[12]This function may be written as: $W = \int \Sigma pq\, dt$

Table I.2: Principles of least action.

	Local effect	Global effect *Extremum function*	Stationariness	Consequences
Geometry	F Element of length	I Length	$\delta I = 0$ Shortest path	Great circle of the sphere
Mechanics	L Lagrangian	S Hamiltonian action	$\delta S = 0$ Hamilton's principle	Law of inertia Principle of relativity
Optics	n Refractive index	N Optical action	$\delta N = 0$ Fermat's principle	Descartes' Law of Refraction

from it will, in the first approximation, make no difference in the action. Any difference will be in the second approximation, if we really have a minimum.

The principle of least action applied to the dynamics of a material point allows us to retrieve Galileo's principle of relativity. In effect, the Lagrange function for a free body, not subjected to an external force, will have the form $L = T$ since $V = 0$. We then deduce that the action S is given by the integral of v^2 between the initial and final instants. According to the principle of least action, the function S, which is at a minimum, corresponds to a constant velocity of the body between the initial and final instants. This corresponds to Newton's first law of dynamics, i.e. the law of inertia, which states that a free movement is rectilinear and uniform, describing a straight line in Euclidean space. Since the Lagrangian remains unchanged from one reference frame to another in which the same rectilinear, uniform movement occurs, the same laws of motion will apply in both reference frames. Therefore, the spatiotemporal properties and the laws of mechanics will also remain unchanged. This corresponds precisely to *Galileo's principle of relativity*.

An important principle that has completely changed our understanding of the universe is Einstein's principle of relativity (Einstein, Lorentz, Weyl and Minkowski, 1952). The consequences of this principle have led to the construction of what must be considered the most beautiful theory in physics today. Einstein's extension of the law of inertia established that "straight lines" belong to a universe possessing a Riemannian metric space.[13] In this space, the "straight lines" are in fact *geodesics*, or curves corresponding to the shortest paths between two points on any surface, taking into account the forces of gravity.

Another example of the principle of least action is to be found in the laws of optics enunciated by Fermat (1601–1665). More precisely, in this case we should refer to the principle of least time since among all the possible paths that light may

---

[13]The Riemannian metric space is given by: $ds^2 = dx^2 + dy^2 + dz^2 - c^2dt^2$, where x, y, z are the coordinates of a point in space, t the time, and c the velocity of light.

take to go from one point to another, the path chosen will be the one requiring the shortest time. A ray of light refracted through different media obviously does not take the shortest path. Fermat therefore suggested that the path taken might be that requiring the shortest time. A century later, the definition of the Lagrange function for this type of system led to the general formulation of the *optical path*, i.e. the path that light would take in a vacuum during the time actually taken. This would be cT, where T is the interval of time and c the velocity of light. Here, the action N is given by the integral of the index of refraction n, which is a continuous function of the position, corresponding to rate of propagation of the phase wave from one point to the other. Fermat's principle can thus be expressed by $\delta N = 0$. From this we can readily deduce Descartes' well-known law of optics: $n \sin i = $ constant, where i is the angle of incidence.

Many other applications of the principle of least action could be cited, for example in electrostatics, electromagnetism, quantum mechanics or relativity. But what is interesting is the power of this principle in situations that may be described as virtual. In the example above, Descartes' law of optics enables us to understand the behavior of light when, traveling in a given medium, it strikes another medium. It is easy to realize that something occurs at the interface, leading to the refraction of light, so that the phenomenon is perfectly understandable. However, with the principle of least time, the understanding becomes problematic. Here is what Feynmann (1969) had to say about the principle of least time:

> The principle of least time is a completely different philosophical principle about the way light works. Instead of saying that it is a causal thing, that when we do one thing, something else happens, and so on, it says this: we set up the situation, and *light* decides which is the shortest time, or the extreme one, and chooses that path. But *what* does it do, *how* does it find out? Does it *smell* the nearby paths, and check them against each other? The answer is, yes, it does, in a way. That is the feature which is, of course, not known in geometrical optics, and which is involved in the idea of *wavelength*; the wavelength tells us

approximately how far away the light must "smell" the path in order to check it.

Such problems can be solved by *variational calculus*. To take a simple example from geometry, a circle is described as the locus of all the points situated at a constant distance from a fixed point. This is the simplest definition, but the circle could be defined in another way. We could say it is a curve of a given length, surrounding the largest surface. In the examples considered above, we looked for the minimum value, whereas here we are interested in the maximum.

It is surely extraordinary that the fundamental laws of physics can be expressed in terms of a principle of least action. True, this principle applies only in the case of conservative systems for which all the forces can be calculated by means of a potential function. However, it should be noted that non-conservative systems exist only at the macroscopic level. In fact, at the microscopic level there are no dissipative systems. Frictional phenomena, for example, appear at the macroscopic level simply because we ignore the underlying microscopic activity of the vast number of particles involved.

The formalization of physical phenomena

There is evidently a close relationship between physics, the science of the material universe, and mathematics. It is certainly remarkable that the intelligibility of nature should be mathematical. How does the mathematical formulation actually work with a scientific construct? First, there is the experiment that allows us to discover the relationships between the results of the measurements. Then, these results are used to define the physical quantities corresponding to mathematical entities. For instance, with Newton's laws of motion, the measurement of the time taken by a sphere to roll down an inclined plane, from one point to another, reveals its *acceleration*, expressed mathematically as a uniform increase in velocity. Here, we can already see how complicated and misleading it might be to express the idea of acceleration in ordinary words. What exactly is a uniform increase in velocity? We could, of course, say that the velocity increases by a certain amount during each interval of time. However, the velocity itself is the variation of the distance during the same interval of time. Again, we might say that the acceleration is the variation of the variation of the distance with time. In fact, we can define acceleration only because a quantified measurement can actually be made. Formalization can then be used to express the acceleration more conveniently. In the physical sense, such variations are mathematically described by differentials, i.e. infinitesimal elements of a variable that, in this case, may be time or space. In fact, it was the physical discovery of the phenomenon of acceleration that led to the development of the corresponding mathematical tools, i.e. differential and integral calculus. However, as we shall see, newly observed physical phenomena can often be introduced within the framework of an established mathematical theory for further investigation.

How exactly do we construct a mathematical formulation for natural phenomena? How does the mathematical process of integration then allow us to make predictions concerning the phenomena observed? Of course, the relationships obtained by the analysis of the phenomena must be reproducible, i.e. measurements made under identical conditions should always have the same values. Thus, in the case of the sphere rolling down an inclined plane, the same measuring instruments, for example, a stopwatch or a pair of photoelectric cells, should indicate the same time intervals. Only reproducible results such as these will allow us to conceive the scientific structure underlying the observation. The first consequence of the existence of a reproducible relationship is that it is valid in all points of time and space. Moreover, this relationship must remain valid even though certain factors, or parameters, may vary with space. In our example, the factor g, the gravitational constant that intervenes in the movement, depends on the place where the experiment is carried out. Between the force acting on a body, located at a given point in space, and the movement of the body from this point to a neighboring point in space, there exists a simple relationship in which the constant of proportionality is the mass of the body, as indicated by Newton's laws of motion. Here we have the secret of the process of integration. This local relationship, existing between two neighboring points, is repeated successively from point to point, and the sum of all the local variations then gives the integrated, global variation between the starting and finishing points of the movement. Metaphorically, we might say that mathematics tells us exactly what will happen a little further in space and an instant later in time since we already know what is happening here and now. There is clearly a close correspondence between the physical phenomenon, i.e. the variation observed in space between two successive instants, and the mathematical solution of the differential equation, since this correspondence exists between the different physical elements, e.g. distance, time, force and mass, and the mathematical symbols of the equation. This one-to-one correspondence may be called *the natural law*, since it allows us to make predictions concerning natural phenomena. According to Ullmo (1969):

> Prediction is possible as soon as there is a parallelism between the variations in the measurements of certain physical quantities and the variations of certain mathematical symbols. This is the principal form of the mathematical expression of natural phenomena. The parallelism, ensured by the constant testing of the reproducibility of results in the physical universe, is imposed by the mathematician.

Here we have the principle underlying the mathematical formulation of a physical phenomenon. We see that it is essentially based upon the superposition of a large number of elementary phenomena, all of which are identical, so that the physically observable global phenomenon corresponds to the mathematical solution of a differential equation.

Once we grasp the mechanism of the elementary phenomenon, the integration of all the elementary phenomena will give us an insight into the complex phenomenon observed. The mathematical formulation is useful because of the simplicity of the elementary phenomenon. For instance, Newton's law of gravitation is simple since it expresses the elementary phenomenon of attraction between two bodies. Again, Coulomb's law expresses the elementary phenomenon of attraction between two electrical charges with opposite signs, and that of repulsion when the signs are the same. When these laws are applied to sets of masses or charges, the mathematical combination of the elementary phenomena becomes extremely complex. However, mathematical reasoning enables us to move from the local to the global aspect of the phenomenon mainly because of certain "natural" properties of non-living matter, such as its *homogeneity* (the physical properties of matter being the same at all points), its *locality* (the effects produced at a distance being negligible), and the *simplicity* of the elementary mechanism. The integration of the elementary mechanisms leads to a higher level of the organization of matter. In this sense, the global phenomenon is explained by the *mathematical integration* of local phenomena.

Evidently, the discovery of new elementary mechanisms is likely to become increasingly difficult as we advance in our knowledge of the physical universe. The three characteristics of non-living matter, i.e. homogeneity, locality and simplicity, may not all be present at a certain level of structural organization. However, certain fundamental principles, such as the conservation of matter and energy, still allow us to describe the physical universe. In contrast, the absence of these characteristics in living organisms may suggest the impossibility of a mathematical representation of biological phenomena. As we shall see in the chapters that follow, biology cannot be reduced merely to physics. The structural organization of the non-living world depends on interactions called forces. In living organisms, such interactions are compounded by a complex set of functional biological interactions that can be physically described. Finally, the combination of these functional interactions produces the self-organization of a biological system.

The formalization of natural phenomena is therefore subject to important constraints. For instance, let us consider the differential equation representing Newton's law: $(d^2x/dt^2 = F/m)$. This is a second-order differential equation with respect to time. However, why do the other derivatives not appear in this equation? Furthermore, why is there no variation with respect to space? It is not easy to answer these questions. In this particular case, experimental observation revealed the local physical variation that is expressed by the differential equation. However, it can be shown mathematically that a second-order differential equation is equivalent to a system of two first-order differential equations. In general terms, a differential equation of order n is equivalent to a system of n first-order differential equations. Such systems are called *dynamic systems*. To return to Newton's law, the mathematical formulation actually introduces a new physical concept — that of the quantity of movement, or *momentum*.

The first question concerns the existence of higher derivatives. In fact, the formulation of certain physical phenomena requires that such terms be introduced into an n-order differential equation corresponding to the dynamic system. The mathematical theory of dynamic systems can then be applied to obtain, if not the solutions themselves, then at least the conditions for the stability of these solutions in the neighborhood of the singular points, i.e. the points of the dynamic system at which the first derivative is null. In other words, in the equation: $dX/dt = f(X)$, we can determine the values of X such that $f(X) = 0$. The behavior of these non-linear differential equations is often rather strange, sometimes producing periodic solutions describing the variation of the physical system with time. Since this behavior depends on the values of the parameters associated with the equations, the variation of the system with time can be mathematically predicted in a way that would otherwise be impossible. Thus, in the case of dissipative chemical structures, subjected to an open thermodynamic regime, the periodic structuring of the medium with time was successfully predicted by solving the dynamic system of equations representing the phenomenon (see Chapter V).

The second question concerns the existence of variations with respect to space. Some equations contain terms, such as dX/dx, d^2X/dx^2, and so on. We then have partial differential equations. As in the case involving variations with respect to time, these equations represent reproducible local, spatiotemporal relationships. In general, partial derivatives of an order higher than the second order play no role in the mathematical representation of physical phenomena.

An example of formalization: the phenomenon of diffusion and the propagation of heat

A classic example of an equation of the type considered above is the reaction-diffusion equation. This equation describes a physical phenomenon in which, during the time dt, the variation of $X(t)$ (for example, the concentration of a chemical solution) is the result of the combined action of two mechanisms: on one hand, the diffusion due to the Brownian motion, or molecular agitation, which is a purely thermodynamic effect,[14] and, on the other, the chemical reaction represented by the function $X(t)$. Since $X(t)$ is a vectorial function, we obtain a system of partial derivative equations.

Although it is rather complex because of the mechanisms involved, this example of the coupling between the reaction and the diffusion is a good illustration of what we have called *the natural law*, i.e. the correspondence between the physical elements and their mathematical symbols. The diffusion process is represented by: $D\nabla^2X$, where D is the diffusion coefficient for the chemical species, the concentration of which in the medium is X, the reaction being represented by $f(X)$. The equation of the variation of X with respect to time may then be written as: $\partial X/\partial t = D\nabla^2X + f(X)$. This equation also shows the locality of the physical variations. In general, this type of equation expresses a physical principle, for instance that of the *conservation of mass* in a small volume situated about a point x between two infinitely close instants: t and $t + dt$. Thus, the formalization is underpinned either by experiments describing the phenomenon through a set of reproducible relationships, or by a few great principles that are sufficiently general to be applicable in a wide range of situations.

---

[14]Einstein (1905) showed that Brownian motion, corresponding to the random movement of small particles suspended in a fluid, could be explained by the effect of the collisions between the molecules of the fluid and the particles.

I

In classic dynamics, the principle of conservation of energy is expressed simply by saying that an isolated dynamical system conserves its energy during its variation with time. The principle may be formulated in terms of the two types of energy known, i.e. the potential energy that depends only on the position of the body, and the kinetic energy that depends only on its velocity. The physical significance of this is far-reaching indeed — the dynamical system varies with time such that the potential energy is transformed into kinetic energy, the sum of the two remaining constant. However, the problem becomes inextricable if we consider a system containing large numbers of particles, since the determination of the dynamics of the system would require the knowledge of the dynamics of each of the particles. For instance, even in a tiny drop of a reactive solution, this might mean analyzing the behavior of 10^{20} particles by solving three times this astronomical number of simultaneous equations — an almost impossible task for the time being.

If this is the case, how do we interpret a physical experiment as simple as that of recording the propagation of heat in a metal bar strongly heated at one end? What we observe in practice is the tendency towards a homogeneous distribution of temperature in the bar, i.e. a thermal equilibrium. According to Fourier (1768–1830), who established the theory of the propagation of heat in solids, the flux of heat between two bodies is proportional to the temperature gradient between them. Mathematically, this is described by an equation containing second-order partial derivatives with respect to space, and first-order partial derivatives with respect to time.

In the case of the metal bar, the solution to this equation corresponds to the distribution of heat according to the well-known statistical law due to Laplace (1749–1827) and Gauss (1777–1855). This remarkable law governs certain random phenomena under very general conditions, as often found *normally* in nature, which is why it is usually called the *normal law of distribution*. The calculation of probability shows that this law applies to a randomly repeated phenomenon, each occurrence of which is independent, and which is subject to the influence of a large number of independent, random factors. For example, if we make repeated measurements of some quantity, under the same conditions, then the values obtained will satisfy the normal law of distribution.

Let us now consider the problem of the random walk, which simulates Brownian motion. In this model, a molecule is supposed to move in a random direction over a unit distance in each unit time-interval. If we calculate the probability of finding the molecule in a given point at a given time, we shall have solved the diffusion equation that, we may recall, expresses a law of conservation of the number of particles. We obtain the same type of solution for the heat equation. The *irreversibility* of the phenomenon of diffusion is thus included in the solution of these models, since the solutions are found to be identical. Although the problem of the propagation of heat and the problem of the random walk involve apparently very different phenomena, the profound analogy between them is expressed by the identity of the solutions obtained. Here we may admire the power of mathematics that has led to the discovery of a fundamental unity that would otherwise have remained hidden.

Mathematics and physical reality

The description of the physical universe from the standpoint of general principles is of such elegant simplicity that it appears truly beautiful — at least from a theoretician's point of view. This is clearly the case for the theory of relativity that showed that the velocity of light is a physical constant, for quantum mechanics that revealed the wave-particle duality of radiation, for statistical mechanics that brought out the remarkable relationship between entropy and probability, to mention only a few of the most striking examples.

The esthetic aspect of a mathematical theory is no less important than that of any great work of art. In both types of endeavor, the subjective beauty we behold springs from our ability to encompass the extraordinary nature of the work in a single glance. Nevertheless, the quality of simplicity we admire is merely apparent. In fact, behind the simplicity of a fugue written by Bach or a Mass composed by Mozart, the shapes and colors of abstract art or the beautifully restrained examples of Romanesque architecture, there lies hidden an extraordinarily complex organization.

The mathematical formulation of a physical phenomenon gives us an immediate insight into its subtle complexity. As our understanding of the physical universe deepens, we keep on discovering new laws based on abstract concepts. Thus, the description of forces acting on elementary particles is far removed from Newton's law since, at this level of description, we have to use the principles of symmetry, the laws of conservation, the mathematical theory of groups and elaborate operators. Nevertheless, the essential simplicity will remain, although a high price may have to be paid through relentless efforts at abstraction. However, in the end we shall be richly rewarded since the mathematical principles involved will surely accentuate our belief in the fundamental harmony of the physical universe.

We may wonder to what extent the abstract world of mathematics might be identified with physical reality. Could the profundity of mathematical concepts actually correspond to some intrinsic truth in our universe? In effect, the mental constructions of mathematicians do appear to have a certain "reality", as if such constructions were more akin to discoveries than to inventions. Nevertheless, we may ask how a new mathematical structure might be discovered through the examination of the initial physical structure. In contrast, the mental construction of reality appears to be sufficiently "natural", leading us to believe in the reality of the mathematical world, i.e. the Platonic world, in the sense of the reality of the physical world, and finally, to believe in the essential harmony between the two worlds. In the chapters that follow, we hope to extend this Platonic view of reality to the biological world. Evidently, this approach is based on pure intuition, or speculation, as some might call it. However, the coherence of the results obtained in biology is such that it could hardly be due to mere chance. Thus, Penrose (1989) claimed that his little finger told him that the human mind was surely more than just a "collection of microscopic wires and circuits", and added:

> Because of the fact that mathematical truths are necessary truths, no real information, in the technical sense of the term, is conveyed to the discoverer. All the information was there all the time. One had simply to put things together to see the answer! This fully agrees with Plato's idea that a discovery (a mathematical one, for instance) is merely a form of *reminiscence*. I have often been struck by the resemblance between the simple fact of being unable to recall someone's name and the inability to find the right mathematical concept. In each of these cases, the concept we looked for is, in a sense, already present in the mind, although in the form of unusual words in the case of a mathematical idea that has not yet been discovered.
>
> If this point of view is to be of use in mathematical communication, we have to imagine that interesting and profound mathematical ideas somehow have a more forceful existence than those that are uninteresting or trivial.

Without going as far as Penrose with regard to the "spirit" of matter, we shall restrict ourselves to the more prosaic "functioning" of matter so as to seek some general principles underlying the "collection of microscopic wires and circuits" sometimes mentioned in connection with the working of the human mind. However, even considering only the concrete aspects of biology, putting aside all notions of the conscious mind, it is rather difficult to do otherwise than believe in the mysterious bond between the Platonic world of mathematics and that of the living world. After all, is there any reason why what is true for physical reality should not also be true for biological reality?

II

THE FUNCTIONAL ORGANIZATION OF LIVING ORGANISMS

Living Organisms Possess a Structural Hierarchy

II

THE FUNCTIONAL ORGANIZATION OF LIVING ORGANISMS

Living Organisms Possess a Structural Hierarchy

Physical science, by definition, is concerned with the properties of non-living matter and energy, and the relationships between them. Biological systems would thus seem to be excluded from the field of physics. However, since living organisms are composed of matter and depend upon energy exchanges to maintain their existence, we may wonder to what extent the fundamental laws of physics might not also apply to biology. Medical science, increasingly linked to the development of the physical sciences, is hardly an independent discipline. Modern medicine, in which the human body is ideally considered from a global point of view, is based on the integration of physiological and anatomical results obtained mainly by physical methods such as microscopy, ultrasound probes, radiological techniques, nuclear magnetic resonance, computer-aided tomography, and so on.

Physiology describes specific biological systems, such as the respiratory, cardiovascular, renal or hormonal systems. Like the extraordinary discovery of blood circulation by Harvey (1578–1657), the understanding of the other physiological systems has been the end-result of much time and effort. However, we still do not know how to describe a *physiological function,* in the *general* sense of the term. A general description could be expected to lead to a global, integrated knowledge of the organism and, in particular, of its development. If we could determine the fundamental concepts underlying the integration of specific physiological functions, then we might be able to discover the general principles governing the physiological stability of living organisms. This is similar to the approach used successfully in physics to establish, for example, the fundamental principles of least action or the conservation of energy and mass. Although a physiological function is generally taken to refer to the functioning of a particular organ, it is actually the result of a great number of processes operating at several different levels ranging from that of the whole organism down to the molecular level. The integrated activity of the physiological functions contributes to one of the essential characteristics of the organism, i.e. its homeostasis, or the maintenance of the stability of the internal medium. However, the comprehension of the phenomenon of homeostasis requires a thorough understanding of the anatomy of the body and the molecular processes occurring within it.

The influence of Galen (c.130–200), who codified the anatomy and physiology of his time, dominated medical science right up to the Renaissance. For centuries, his work remained the uncontested reference in medicine in spite of the innumerable errors and unsubstantiated statements it contained. It was not until the sixteenth century that these were finally refuted by Vesalius (1514–1564) in his revolutionary work: *De humani corporis fabrica libri septem.* Although surgery was probably practiced even very long ago, as attested by the finding of trepanned skulls dating back to prehistoric times, a real body of anatomical knowledge began to be gradually assembled only when it became possible to borrow appropriate techniques from other branches of science. Thus, up to the nineteenth century, the main obstacles to progress in anatomy were due to pain and infection. Then, Wells (1815–1848) first used nitrous oxide as an anesthetic to combat pain in dental surgery, while Semmelweiss (1818–1865) discovered that puerperal fever could be avoided by antisepsis and asepsis. Rapid advances in pharmacology and microbiology followed the pioneering work done by Pasteur (1822–1895). The terrible casualties of the two World Wars in the twentieth century led to the extraordinary development of the techniques of organ transplants that would have been impossible without the understanding of immunology. These complex operations involved so much medical teamwork that no one name can really be singled out for credit. History gives innumerable examples of this kind in all fields of medicine.

The functioning of living organisms is based on chemical reactions. Physiological mechanisms depend on the integrated activity of the bodily structures at various levels of organization, from molecules to cellular organelles, from cells to tissues, from tissues to organs, from organs to systems, and finally to the whole organism in its environment. How do we define the passage from the chemical reaction to the physiological function? As we shall see, physiological activity actually results from *structural discontinuities*. The systematic chemical organization of a living organism gives it a completely different status from that of non-living matter. We propose to consider a few examples[15] at different levels of organization, e.g. molecular genetics; energy metabolism at the cellular level; cognition and mental representation based on neuronal networks; and some of the major functions of the organism, such as respiration and blood circulation. We shall also discuss heart failure, caused for instance by cardiac insufficiency following arterial occlusion, as a typical example of a pathology stemming from the perturbation of an integrated set of physiological functions.

The functional organization of biochemical pathways

It is often said that a living organism is just a complex, coordinated set of chemical reactions. This is a considerable

---

[15]However, the descriptions given here will be rather brief, our objective being to bring out the main elements. Some details are given in the paragraphs in small print and in the footnotes, but readers are invited to look up some of the works cited in reference for a more comprehensive treatment of the topics mentioned.

oversimplification since the body is evidently much more than a mere chemical factory. What is it that distinguishes the chemical reactions taking place in a test tube from those occurring in living cells and cellular structures?

What exactly is a chemical reaction? The combination of molecules in a material medium involves two successive operations. First, the molecules are transported and brought into contact with each other, and then they share some of the peripheral electrons situated on the outermost layers. The first operation is due to diffusion (see Chapter I), i.e. the random motion of molecules in a constant state of agitation leading to collisions. The intensity of the collision depends on the energy of the molecules, which is determined by the temperature of the medium containing the molecules. The second operation consists of crossing the potential barrier represented by the *activation energy* of the molecules. This term, from quantum physics, describes the mechanism that enables two molecules to combine such that the electron cloud of molecule A penetrates that of molecule B. For this to happen, molecule A must have the energy required to overcome the energy barrier of molecule B. The action of such forces of interpenetration can be explained in terms of the potential energy of the molecular system.

A molecule is a system composed of atoms with electrostatic interactions. The nucleus of each atom vibrates and spins about its axis. The total energy of the molecule is the resultant of the energies of vibration of the nuclei of its atoms, the energies of the individual electrons, and the potential energy of interaction between the atoms. This potential energy is called the *potential reaction surface* since a point of this surface moves during the reaction. For simplicity, let us consider the case of a molecule with only two atoms. At an infinite distance, the atoms would of course be independent. Now let us imagine that the atoms come closer together. In the absence of any interaction, the potential energy will remain constant, being equal to the sum of the individual potential energies of the atoms. However, at a certain distance, an interaction will occur between the atoms such that their total potential energy decreases because of electrostatic attraction. This phenomenon is similar to that of gravitational attraction because of which, when a body falls to the ground, its potential energy decreases while its kinetic energy increases. The comparison can be carried no further, for in the case of the atoms, as the distance between them decreases, there is a region of transition[16] in which the attraction is balanced by the electrostatic repulsion. If the distance between the atoms is then further decreased, the potential energy will increase asymptotically to infinity since the atoms cannot interpenetrate. Thus, mathematically speaking, the potential function constitutes a domain of attraction in which the lowest point corresponds to the equilibrium distance between the atoms where the electrostatic attraction is equal to the repulsion. This is the point of the stability of the system. The energy required to bring atoms, from an infinite distance to this equilibrium distance is called the *dissociation energy* or the *binding energy*. The two atoms can only be united if this *potential barrier* is overcome. The reasoning above could be extended to the more general case of the interaction between molecules containing any number of atoms but the potential surface will of course be highly complicated, making it extremely difficult to represent the pathways of the chemical reactions.

It is well known that heating a mixture, or stirring it, or even exposing it to light or some other radiation, will provide the energy required to produce the chemical reaction between the molecules it contains. Thus, an increase in the temperature leads to increased diffusion that, in turn, overcomes the potential barrier between the molecules. *By its very nature, a chemical reaction is a local phenomenon.* So how do living organisms, in which there is no such heating or stirring, or exposition to radiation, solve the problem? *The explanation is that living organisms contain structural discontinuities between which functional interactions take place.* Here, there are no forces acting at a distance, as in the

---

[16]This transition region is determined by the *distance of interatomic equilibrium.*

case of non-living matter, since such forces can only operate when the medium is homogeneous. Since information has to be transmitted between the physical structures of living organisms, *the chemical pathways are governed by the principle of non-locality.*

Let us consider, for example, the functioning of the thyroid gland. The action of the thyroid on, say, a muscle cell in the leg will require a *non-local* interaction carried out by a vector — in this case a hormone called *thyroxin* — that will trigger the appropriate chemical reaction at a distance. Similarly, when there is an excess of glucose in the blood, the pancreas will increase its secretion of insulin, which in turn will increase the glucose uptake of distant liver and muscle cells. Such non-local interactions occur in all the structures of the organism. Thus, the immense network of chemical pathways in a biological organism corresponds to a network of non-local functional interactions operating between local chemical reactions, as shown in Figure II.1.

> Thus, if there is a set of highly regulated local chemical reactions in an organism then there must also exist non-local interactions between structures situated at a distance in physical space.

The non-locality has its origin in *structural discontinuities* and, consequently, in the global functional regulation of the organism, which must act as a whole. This observation has profound theoretical implications. It allows us to distinguish living organisms from physicochemical systems by introducing the notion of a *function*. In the example above, the function of the thyroid is to act on a cell situated at a distance in physical space, this action being coordinated with other phenomena, obviously of chemical origin, in the organism. Of course, this raises the problem of how the body actually determines the "best" organization for its functions. In Chapter V, which addresses this question in detail, we shall see that the properties of non-locality and structural hierarchy being closely linked, the "best" organization would seem to be a hierarchical organization. The examples that follow should illustrate this idea.

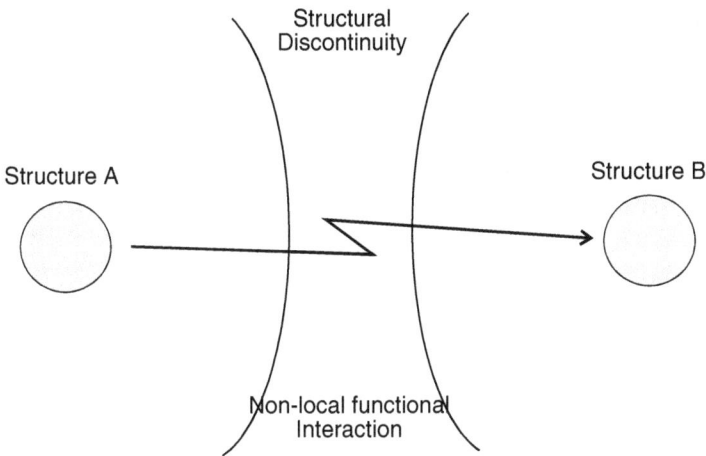

Figure II.1: A schematic chemical pathway in a living organism. The functional interaction is produced at a distance between the two structures A and B, separated by a structural discontinuity. Local chemical reactions occur in each of the structures A and B. If these structures contain no discontinuities, the local reactions may be assimilated to reactions taking place in a homogeneous medium similar to that of a test tube.

A striking example of biological organization is given by the enzyme-substrate system that greatly accelerates the chemical reactions[17] in the body. The system operates at a specific, highly localized site, called the *active site.* The reaction product is then "exported" to act at a distance on other structures of the body. It is difficult to conceive of a living organism based on some other principle since this would most probably lead to an inanimate object, i.e. a homogeneous structure in which the phenomena would all be local. The enzyme allows the execution of one of the main processes of a chemical reaction: that of crossing the potential barrier without any increase in temperature. Two molecules, such as the ligand and the substrate, each with a very low potential energy, will not be able to

---

[17]The kinetics of this chemical reaction, which may be written: $E + S \rightleftharpoons ES \rightleftharpoons E + P$, is given by the Henri-Michaelis-Menten equations. Under certain specific conditions, the solution of the dynamic system of equations gives the rate of production of P.

combine even if they enter into contact by diffusion. However, in the presence of the enzyme, they form an activated complex, i.e. a complex in which the stability of the bonds is sufficiently weakened to overcome the potential barrier. Once a chemical bond is established between the two molecules, the enzyme returns to its initial state, ready for further catalytic activity. The active site, at which the catalytic reaction occurs, contributes to speeding up the process by a factor of at least one million. For example, carbonic anhydrase, an enzyme that facilitates the transfer of carbon dioxide from the tissues into the blood, and then from the blood into the air in the alveoli of the lungs, combines carbon dioxide molecules together millions of times per second — a reaction that would be impossible at ambient temperatures. This enzyme reaction, which is about 10 million times faster than a similar, non-catalyzed reaction, is one of the fastest enzyme reactions known. Of course, the quantity of the product synthesized by the chemical network has to be controlled to satisfy the demand of the body. This involves the action of another network of biochemical pathways. How is the enzyme-substrate system regulated? This can be done in at least two ways. One way of doing this is by controlling the quantity of enzyme synthesized by means of the *epigenetic* system described below; and the other is by activating or inhibiting enzymatic activity by direct or indirect interactions, called *allosteric interactions.* Thus, in addition to its catalytic activity, the enzyme possesses its own regulatory function.

The enzyme-substrate complex is located at a *specific active site*, generally situated at a distance from the point of emission of the substrate. It thus acts as the effector, or the *sink,* of a functional interaction. However, although the concept of the active site is very suggestive, it raises the question of how the enzyme and the substrate actually come into contact. The idea that immediately comes to mind is that this could be due to a random collision as in the case of gas molecules. However, since the collision between two molecules is not very frequent, it is difficult to conceive of a ternary collision, occurring more or less simultaneously, without considering the possible action of some adjusting mechanism. Static adjustment would require that the atomic groups of the enzyme be precisely oriented with respect to the corresponding atomic groups of the substrate. Koshland (Koshland, Nemethy and Filmer, 1966) suggested a mechanism of induced adjustment in which a change in the geometrical structure of the enzyme is followed by the enzyme-induced structural modification of the substrate. The difficulties involved in such geometrical adjustments between the molecular "surfaces" of enzyme and substrate are further compounded by the thermodynamic problems that arise. These geometrical considerations form the basis of the allosteric phenomena described by Monod, Wyman and Changeux (1965).[18]

The current knowledge of enzymatic processes indicates that a non-local[5] functional chemical interaction takes place at an *active site*, involving two highly complex mechanisms that appear to be fundamental to the chemistry of all living organisms, i.e. the *allosteric interaction* and *enzymatic catalysis*. Since each chemical reaction is catalyzed, in principle, by a specific *enzyme*, the number of enzymes should match the number of chemical reactions. The non-local interaction solves the problem of the regulation of a "local" set of chemical reactions, which may be thought of as constituting a homogeneous medium. Let us consider an example.

---

[18]The models most frequently used for allosteric transitions are due to Monod, Wyman and Changeux (1965) and Koshland, Nemethy and Filmer (1966). Theoretically, it is possible — though rather difficult — to describe the kinetics of enzymatic reactions. One method is based on the calculation of the number of possible states of an enzyme-substrate system in a given configuration by means of the *binding polynomial* (Wyman, 1964). In spite of certain limitations, this formalism offers the physical interpretation of *cooperativity*, i.e. a phenomenon in which the dependence of certain sites of the enzyme-substrate system increases (amplification), or decreases (inhibition), according to the activity of the enzyme. It also allows us to calculate the *saturation function*, which is proportional to the number of sites occupied with respect to the total number of sites. Applied to the Monod, Wyman and Changeux model, this function gives an interpretation of the sigmoidal form of the curve representing the fixation of oxygen by hemoglobin, and which can consequently be used to predict the behavior of a patient in respiratory distress from the partial pressure of oxygen, among other parameters.

II

The cells of living organisms obtain the energy they need from organic molecules and molecules of oxygen. The former are provided by the digestive process, which breaks down the food absorbed by the organism, and the latter by the respiration. However, the actual production of energy involves several complex biochemical pathways in specialized organs. *Energy metabolism* is not only a product of the organism but also contributes to its functioning.[19] This metabolism is regulated by the activity of two hormones: *insulin*, the hormone of abundance; and *glucagon*, the hormone of insufficiency. Further details could be found in standard works on biochemistry (see, for example, Stryer, 1992).

Just as interactions between chemical reactions produce chemical networks, those between intracellular and supracellular structures give rise to inter-network interactions. Some of the reactions may be specific, as in hepatic cells, the main function of which is the chemical decomposition of molecules. In other cases, the reactions may be common to all the cells in which they ensure the metabolism. Although such *networks* may be *local* for certain structures, they are no longer so when we consider the corresponding physiological functional system. Thus, the operation of the three fundamental systems of the cell, i.e. energy production, protein synthesis and ion transport, depends on biochemical mechanisms involving various cellular structures, such as the nucleus, the mitochondria, the microfilaments, and so on. Again, such networks can no longer be considered as local when we envisage supracellular structures.

---

[19]The integrated regulation of the metabolism involves two concepts. One is the *dynamic equilibrium between anabolism and catabolism*, the former consuming energy and the latter providing energy. The other is the *molecular convertibility between proteins, hydrocarbons and lipids, which characterizes the passage between the two principal, non-symmetrical states of the organism, i.e. the fasting state and the postprandial state.* In the latter, the abundance of glucose directly provides the energy needed through the glycolytic pathway. The surplus is stored by lipogenesis, i.e. the formation of fats, and glycogenogenesis, i.e. the formation of glycogen. In the fasting state, the lack of glucose leads to an inversion of the metabolic process so that the pathways used are those of gluconeogenesis and ketogenesis, i.e. the formation of ketones.

If the structures labeled A and B in Figure II.1 contain structural discontinuities, then a hierarchy of groups of functional interactions should appear. In short, one or more schemas, identical to the formal schema in Figure II.1, should appear in the structure. The non-locality is, in fact, a counterpart of the hierarchy. Let us illustrate the fundamental relationship between the non-locality and the hierarchy using the example of energy metabolism, in particular, that of *glycolysis*, or the breaking down of glucose, which is the main source of energy in living organisms (Goldbeter, 1974, 1977). We shall then generalize the concept to the hierarchy of the full set of chemical networks in the body.

Glycolysis is a beautiful example of interactions between intracellular structures. The object of the transformations leading from glucose to carbon dioxide is the production of molecules of adenosine triphosphate (ATP), which contain the energy required for all the chemical reactions of the organism. Krebs (1900–1981) discovered the 11-step cycle of reactions by which sugars, fatty acids and amino acids are metabolized to yield carbon dioxide, water and the high-energy phosphate bonds that transform adenosine diphosphate (ADP) into energy-rich molecules of ATP. The ATP molecules are then broken down into ADP, releasing the energy required by chemical reactions to overcome the potential barrier. This is a reversible reaction since ATP molecules can be reassembled from ADP as long as energy is available. Finally, the energy consumed by the organism comes from the environment, more precisely from a link in the food chain. All the transformations of Krebs' cycle take place within the mitochondria; organelles that occupy up to a quarter of the total volume of the average cell. The mitochondrion, in which oxygen is ultimately used in chemical reactions that allow the assembly of ATP molecules, may thus be viewed as the true chemical effector of the so-called "silent", cellular respiration (Baillet and Nortier, 1992).

The glycolytic pathway leading to ATP synthesis in mitochondria has been extensively investigated, theoretically and experimentally, *in vivo* as well as *in vitro*, in yeast cells and in red blood cells (Heinrich and Rapport, 1976). We now know that the physiological operation of the ATP subsystem may be affected by the end products, such as pyruvate, or by various metabolites from other pathways. The activity of the subsystem is coordinated by general mechanisms as well as specific mechanisms of local regulation (Queiroz, 1979). This regulation involves not only glycolysis but also several other associated processes[20] (Hess and Boiteux, 1971; Goldbeter and Lefever, 1972).

Since all cells are fueled by glucose, which is transported through the body, the production and the consumption of glucose has to be regulated. This regulation is due to the action of two hormones, i.e. insulin and glucagon. *Thus, the glycolytic process occurring at the intracellular level is regulated at the supracellular level.* What is even more remarkable is the extraordinary *specialization* of the cells involved in energy metabolism. Glucose is stored mainly in muscle cells and fat cells, whereas much of the consumption takes place in brain cells, which use up to 20% of the total energy produced by the body. Liver cells transform glucose according to the needs of the body. These cells, each of which contains between 1000 and 2000 mitochondria, ensure the vital energy functions of the organism. They capture glucose, store reserves, transform the surplus into fatty acids, break down amino acids, and synthesize several proteins. We may recall that investigations in this field were pioneered by Claude Bernard (1813–1878) who demonstrated the role of the liver in the synthesis and breakdown of metabolites. His discovery of the glycogenic function of the liver introduced the important notion of the internal secretion of the body.

Thus, cellular specialization, an essential aspect of the higher living organisms, is represented by the location of specific organs in different parts of the body. However, would

------

[20]It is interesting to note that the addition of a sugar, fructose-6-phosphate, or metabolites situated at the upper end of the chain, has been experimentally shown to trigger *glycolytic oscillations* in the concentrations of all the metabolites of the chain.

specialization be possible at all without this spatial organization? Of course, this disposition of specialized tissues and organs implies the complex transport of metabolic products, which has to be regulated at the level of the whole body.

In effect, the autonomous living organism has to satisfy the principles of homeostasis, i.e. the biological parameters, such as the plasma concentrations of calcium or glucose, have to be maintained reasonably constant. From the physical point of view, the organism should remain in a stationary, non-equilibrium state. From the physiological point of view, if there should be some deviation from the *functional* state of equilibrium, then the regulatory mechanisms of the body should return it to its initial functional state as soon as possible. For example, the level of glucose in the blood, or *glycemia,* should remain constant in spite of the daily variations of the glucose uptake. This regulation depends on the activity of six hormones, only one of which — insulin — reduces the glucose level, whereas the other five, including glucagon, have the opposite effect. After a meal, the food absorbed is broken down in the digestive tube. Venous blood from the digestive tube carries the broken-down molecules directly to the liver through the portal vein, bypassing the general blood circulation. In the liver, a network of biochemical pathways, the complexity of which is only beginning to be understood, transforms most of these molecules. The sugars then enter the bloodstream producing hyperglycemia, which is quickly corrected by the insulin secreted by the pancreatic cells that instantly recognize the "abnormal" level of sugar in the blood. Transported by capillaries in contact with the cells of the body, the insulin increases glucose transport across the cell membranes by up to 500%. As we have seen above, the glucose that is taken up by a cell is used either by glycolysis for its immediate energetic needs, or saved for eventual use in the form of glycogen, or stored in the form of fats.

The example of glucose regulation allows us to identify the three main elements of the functional interaction, i.e. the *source*, the *functional interaction* and the *sink*. Here, the *source* is a pancreatic cell, or more precisely a β-cell in one of the Langerhan's islets of the pancreas. The source is the site of several transformations, from preproinsulin to proinsulin, and then from proinsulin to insulin.[21] The vector of the *functional interaction* is the insulin molecule,[22] which is transported to all parts of the body by the bloodstream. During the transport, the insulin molecule binds closely to a receptor — the *sink* — situated on the plasma membrane of the target cells.[23]

Thus, glucose regulation involves a complex set of chemical reactions occurring in structures located at a distance from one another in the body. This is something quite different from what happens in a homogeneous medium such as a solution in a test tube. The functional interactions between the isolated structures of the body are, by very their nature, non-local.

Can the example of glycolysis, as considered above, be generalized to give an abstract representation of metabolic pathways and

---

[21]The proinsulin is transported through the sacs of the Golgi apparatus of the β-cells of the pancreas before being enveloped in a membrane to form a secretory vesicle. At this stage, a proteolytic enzyme — a protease — hydrolyzes the polypeptide chain of proinsulin to produce the insulin molecule. The secretory vesicle, in which the insulin is stored, is then transported as required towards the cell membrane where membrane fusion releases the insulin into the bloodstream.

[22]Insulin secretion, first rapid and transitory, is followed by a slower, prolonged phase. Insulin circulates freely in blood plasma but it is rapidly broken down in the liver, its half-life being only about five minutes. Insulin production is an oscillatory phenomenon with a period of about 10 minutes.

[23]Insulin has a high affinity for its receptors, and as the concentration of receptors is very weak, i.e. about one receptor per μm^2 of cell membrane surface, corresponding to 10^2 to 10^5 receptors per cell, the concentration of insulin in the blood is extremely low (about 0.1 nM). From the point of view of the functional interaction, the sink consists of a set of receptors on the cell membrane. The receptor, which is an enzyme, catalyzes the phosphorylation of the tyrosine residues of the target proteins. This triggers intracellular tyrosine kinase activity that inhibits catabolism and stimulates anabolism, causing the cell to take up glucose from the blood. This mechanism has three consequences. First, the specific enzymes that differ according to the target cells, e.g. adipocytes and muscle cells, are activated. Secondly, the glucose transporters in the cell membrane are translocated, for instance, *via* insulin-sensitive transmembrane canals. Finally, the transcription of genes corresponding to certain enzymes, such as glucokinase and pyruvate kinase, is increased leading to a stimulation of anabolism and accelerated cell growth.

of integrated energy metabolism? Local biochemical pathways have already been schematized and thoroughly investigated, experimentally as well as theoretically (Savageau, 1974; Savageau, 1979). How do we represent the set of non-local functional interactions involved in energy metabolism? In the case of glycolysis, we identify, on one hand, the insulin produced by the β-cells of the Langerhan's islets in the pancreas, and on the other, the set of targets of the secretion, i.e. all the cells of the body. Figure II.2, which corresponds to the formal schema in Figure II.1, gives an idea of the functional organization of energy metabolism.

As we shall see, several other interactions related to energy metabolism, for example those taking place in the cell nucleus, have already been identified. However, for the present, let us just consider the local kinetics of each of the structures shown, with transport from one to the other, i.e. from source to sink, satisfying the formal schema of the functional interaction shown in Figure II.1. Since the physical structures are arranged in the form of compartments within compartments, with interactions occurring at a distance, it is easy to see how the functional organization corresponds to a *hierarchical system*. In the next section, we shall consider the functional organization of the gene in a similar way.

The functional organization of genes

Gene structure and function have been treated in considerable detail by several authors (Watson, 1978; Stryer, 1992). Here, we propose merely to recall some of the essential elements and stress some of the points that might help us to appreciate the hierarchical functional organization of genes.

What is a gene? The definition of a gene has changed considerably since the early work on heredity, mainly because of the extraordinary scientific and technological progress that has been made in molecular biology after the pioneering genetic experiments carried out by Mendel (1822–1884). Gros (1986) gives a good account of the fascinating developments in this field. Shortly after Mendel (1866) had published his remarkable findings on the hybridization of green peas, Flemming (1882)

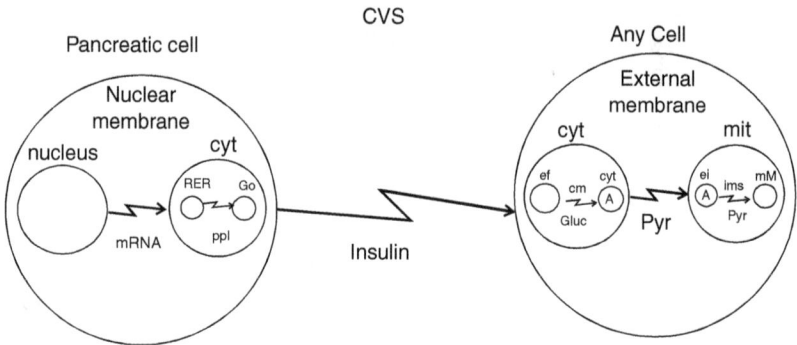

CVS

Pancreatic cell

Any Cell

Nuclear
membrane

External
membrane

nucleus

cyt

cyt   mit

RER   Go

ef   cyt
cm   A
Gluc

ei   ims   mM
A
Pyr

mRNA   ppl

Pyr

Insulin

Figure II.2: Energy metabolism viewed as a hierarchical system operating through functional interactions. (a) *The endocrine secretion — insulin — is produced by the β-cells of the Langerhan's islets in the pancreas.* The messenger RNA (mRNA) is transported from the nucleus, through the nuclear membrane, to the cytoplasm (*cyt*). In the cytoplasm, the mRNA promotes the synthesis of preproinsulin (*ppI*) within the rough endoplasmic reticulum (*RER*), from where it is transported across the cytoplasmic space to the Golgi apparatus (*Go*). Here, the preproinsulin is transformed into proinsulin and released, through the cell membrane, into the cardiovascular system (*CVS*) where it enters into the blood plasma. (*b*) *The set of insulin targets* is composed of all the cells in the body. The insulin binds to specific receptors situated on the external face (*ef*) of the cell membrane. This triggers a series of events: glucose channels allow greater quantities of glucose to pass through the cell membrane (*cm*) into the cytoplasm, where the glucose undergoes anaerobic glycolysis, producing pyruvate. The pyruvate penetrates the outer mitochondrial membrane (*omm*), passing first into the intermembrane space (*ims*) and then, across the inner mitochondrial membrane (*imm*), into the mitochondrial matrix (*mM*). Here, the pyruvate triggers a series of transformations (the Krebs cycle), leading to the production of ATP. A part of the hierarchical system of energy metabolism is represented here in terms of the functional interactions between source and sink. The hierarchical levels of these functional interactions, in descending order, correspond to those of *insulin, mRNA, pyruvate, glucose and preproinsulin.*

discovered dense filaments in cell nuclei that he called chromatin. Then Waldeyer (1902) observed the characteristic figures formed by chromatin in dividing cells. These structures, called chromosomes, gave rise to considerable speculation in genetics, and finally Morgan (1910), with his experiments on the fruit fly, *Drosophila melanogaster*, established the chromosomal theory of heredity. These experiments were based on mutations, i.e. the sudden change in the state of a gene brought about by some chemical or physical agent. The analysis of the order and the spacing of genes enabled Morgan to construct a *chromosomal map*, demonstrating that hereditary characters do not merely mingle but recombine. The calculation of the probability of gene recombination led to the development of what might now be called *gene topology*. Finally, Müller (1927), using the technique of X-ray mutagenesis, showed that the gene could actually be considered as the *unit of mutation and recombination.*

How do genes work? It was far from easy to establish the connection between the active gene and its physical expression in an organism. The first step was to recognize that the biochemical products of the body are in fact the result of gene expression. Thus, Ephrussi and Beadle (1935) showed that the eye color of the fruit fly depends on the activity of two hormones that are produced by specific genes. As a result, several years before the development of molecular biology, this discovery opened up the field of *physiological genetics*. The new analytical techniques invented in the first half of the twentieth century demonstrated that almost all the metabolites known were elaborated by similar biochemical pathways, schematically represented in Figure II.3. These findings prompted Beadle and Tatum (1941) to formulate their well-known aphorism: "one gene — one enzyme". However, it was soon observed that if a mutation occurred in the structure of a gene, the corresponding enzyme would either not be synthesized or, if it were, it would generally be *inactive*. Thus, a defect in one of the subunits of the gene was enough to render the enzyme inactive, and lead to the production of an abnormal metabolite at the end of a metabolic pathway. This phenomenon is illustrated by the sickle cell disease, which is caused by the appearance of a single point mutation in the hemoglobin gene. Thus, Beadle and Tatum's aphorism might be more precisely

Genes G_1 G_2 G_3

mRNA R_1 R_2 R_3

Enzymes E_1 E_2 E_3

Substrates $S_0 \longrightarrow P_1 \longrightarrow P_2 \longrightarrow P_N$

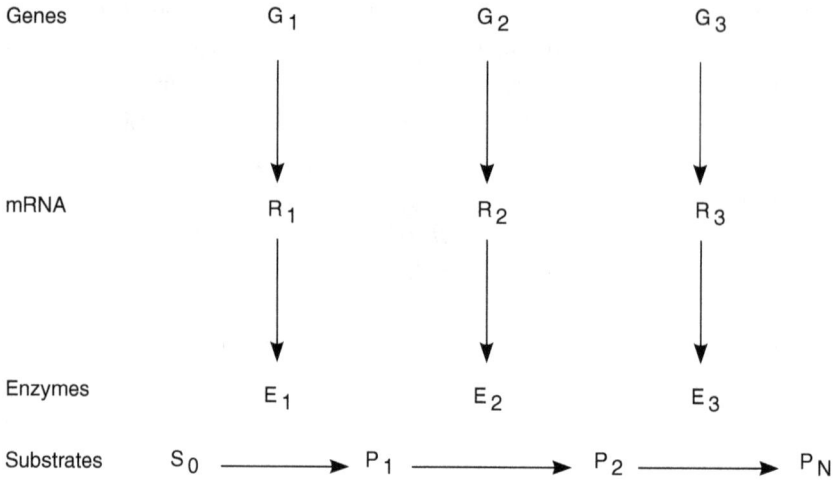

Figure II.3: The epigenetic and metabolic systems. Specific enzymes, synthesized by the epigenetic system, catalyze the chemical reactions in a metabolic pathway. For example, activating structural gene G_2 leads to the translation of the appropriate mRNA R_2 to produce enzyme E_2. Finally, enzyme E_2 transforms substrate P_1 into substrate P_2 in the metabolic pathway.

stated as: "one gene — one polypeptide chain". *The gene could then be considered as a unit of mutation, recombination and function.*

Unfortunately, it quickly became clear that the apparent simplicity of this idea masked a rather more complex reality. New experimental results in the field of genetics quickly led to the invention of neologisms, such as "cistrons", "recons" and "mutons".[24] These terms were coined by Benzer (1957) along the same lines as the words used by physicists to describe the new elementary particles then being discovered, e.g. kaons,

---

[24]The *cistron* is the smallest unit of genetic material that can transmit genetic information, i.e. determine the sequence of amino acids of a specific polypeptide chain. The *recon* is the smallest unit of genetic material capable of recombination, presumably comprising a set of three nucleotide bases, or a triplet. The *muton* is the smallest element of genetic material that, once altered, gives rise to a mutant form of the organism.

muons, pions, and so on. Interestingly enough, Benzer was able to predict that the real distance between two recons, i.e. between two sites at which mutations occurred, would be of the same order as that separating two nucleotides. Thus, chromosomal structure was analyzed, in a sense, at the molecular level even though nothing was then known about the nature of the polypeptides or even of the chemical sequence of their constituents (Gros, 1986).

Let us now consider the structure of the filament found in the chromosome, and later identified as deoxyribonucleic acid (DNA). At the time of its discovery, the structure was believed to be of a proteinic nature, given the powerful, specific catalytic properties then attributed to proteins. Then, following the work done by Griffith (1922) on the *pneumococcus*, the bacillus that causes pneumonia, Avery, MacLeod and McCarty (1944) described a "transforming principle" that could turn non-infectious cells into virulent cells. The virulence had been found to depend on a reversible gene mutation. The chemical purification of this so-called "principle" revealed that it was in fact not a protein at all but a highly polymerized, viscous form of deoxyribonucleic acid. However, when it was suggested that the chromosome filament might be composed of simple molecules of nucleic acid, the idea found little favor. Finally, Watson and Crick (1953) revealed the molecular structure of deoxyribonucleic acid, that in a way substantiated the original idea of the "transforming principle", which was finally identified as a mixture of bacterial chromosome fragments made up of DNA.

Watson and Crick's discovery heralded the extraordinary development of molecular biology. Jacob and Monod (1961) were able to explain their concept of *reproductive invariance*, i.e. the exact reproduction and multiplication of highly organized structures, based on the structure of the DNA double helix. Thus, an answer to the complex problem of heredity was given by a relatively simple molecular mechanism. The idea was that reproductive invariance from one cell generation to the next depends on the structural symmetry of DNA, and that this symmetry is maintained by the principle of nucleotide base pairing. Base pairing was experimentally demonstrated in DNA replication in *Escherichia coli* by Meselson and Stahl (1958), and the results were extended to the three main steps of protein synthesis, i.e. *replication, transcription* and *translation*, briefly described below.

> *Replication*, which allows the "perfect" reproduction of the DNA molecule during cell growth and division, *takes place in the cell nucleus*. It is based on the structural complementarity of two pairs of nucleotide bases, adenine and thymine, on one hand, and guanine and cytosine, on the other. The assembly of the double helix of the DNA involves specific enzyme activity. During replication, the two strands of the double helix are separated, and each strand serves as a matrix for the formation of a new double strand, replicating the original portion of DNA. The process is repeated all along the length of the molecule, finally resulting in the total duplication of the DNA preparatory to cell division.
>
> *Transcription* is the process by which a gene, inscribed in a nucleotide sequence on one of the strands of the DNA molecule, is copied in the form of a single-stranded RNA molecule. The copies are made according to the same nucleotide pairing rules as for replication. *Transcription, like replication, takes place in the nucleus*. According to the gene that is copied, three types of RNA are formed: messenger RNA (mRNA), transfer RNA (tRNA) and ribosomal RNA (rRNA). The RNA molecules then migrate across the nuclear membrane into the cytoplasm. In the higher organisms, the transcription involves a "maturation" stage during which the copies of certain non-coding DNA sequences are eliminated.
>
> *Translation* corresponds to the biosynthesis of proteins, i.e. the assembly of amino acids in the order specified by the nucleotide sequence of the mRNA.

The chemical phenomena underlying protein biosynthesis consist of local processes of reaction-diffusion taking place in *different* intracellular structures. We may therefore identify the functional interaction with the migration of messenger RNA from the nucleus, through the nuclear membrane, into the cytoplasm. Here, the nuclear membrane represents the structural discontinuity involved in the functional interaction. The schema in Figure II.4 represents the functional organization of the synthesis of a protein.

It should be observed that the flow of information is non-local. The messenger RNA has to be transported from a source in the nucleus, across the nuclear membrane, to a sink in the cytoplasm. The compartmentalization of the processes of transcription and translation is probably linked to the specialization of cellular structures, and consequently to that of the corresponding functions. It is difficult to imagine that the routine operations involved in the everyday, physiological functioning of a cell, i.e. the synthesis of proteins, could be carried out within the same space as the exceptional process of DNA replication, essential to the reproduction of the cell. It is true that this argument has a somewhat finalistic quality but, as we shall see in Chapter V, the principle already proposed in the case of energy metabolism may be generalized to cover the flow of information.

How does the gene interact with its medium? The structure of DNA explains the conservation of heredity, since it allows the synthesis of the enzymes required for the biochemical reactions leading to the end product, and finally to the maintenance of the phenotype. In particular, what causes certain genes to be switched on while others are switched off? This

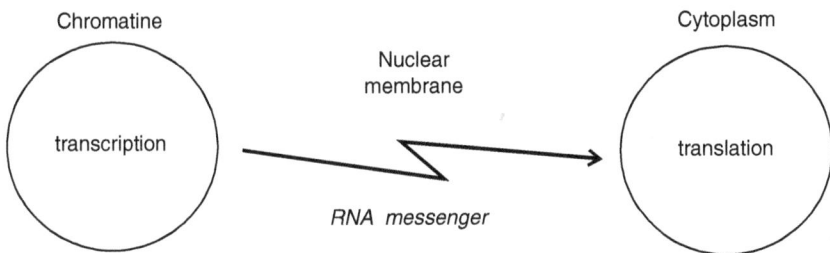

Figure II.4: Representation of protein synthesis following the formal schema shown in Figure II.1. The functional interaction here corresponds to the action of the mRNA molecule.

problem was investigated in prokaryotes, and the results obtained were shown to be so general that the cybernetics of the processes could be extended to all living organisms. The main point that emerged was that the gene is in fact activated or inhibited by the cellular medium itself.

Among the prokaryote genes, Jacob and Monod (1961) identified certain independent groups of genes they called *operons.* The operons consist of closely linked structural genes the activity of which is controlled by an *operator* gene through its interaction with a *regulator* gene. In bacteria, this regulation depends upon the concentration of certain metabolites, called *inducers,* which are not used in any specific biochemical pathway.

> If the inducer appears to be in excess, it will no longer be synthesized. In other words, the gene responsible for its synthesis will have to be repressed. On the contrary, if there should be a shortage of the inducer, its synthesis would have to be speeded up. The unused portion of the inducer binds to an aporepressor, i.e. the inactive form of the repressor produced by the regulator gene. This binding transforms the aporepressor into an active repressor that binds to the operator gene, causing it to block the structural genes, thus stopping the synthesis of the corresponding proteins. The expression of the structural genes will be resumed only when the repressor and the operator are separated.

Thus, we have a functional system consisting of three elements, i.e. the operon, the repressor and the inducer. Two situations may be envisaged: either the inducer inactivates the repressor, so that the induction corresponds to a derepression since it cancels an inhibitory effect; or the enzyme combines with the product of a regulator gene causing the activation of the operon. The effect produced may be considered "positive" in the former case, and "negative" in the latter. Whatever the actual process, *a molecule (the repressor) emitted by a source (the regulator gene) acts at a distance on another molecular structure (the operon), thereby transforming its properties.* This corresponds to the functional interaction as defined above. However, we do not know the precise distance over which the action occurs, particularly because of the complex way the DNA is coiled around the histone molecules. Nor do we know whether the processes involved are local or non-local. Again, is molecular movement due strictly to diffusion forces or is it produced by a mechanism of transport, as in the case of the mRNA molecule? For the time being, we shall assume that the molecules playing a role in the mechanism of regulation, i.e. the structural genes, the operator gene and the regulator gene, participate in local processes.[25]

Gene dynamics was revolutionized when McClintock (1952) showed that intense physical perturbations produce unstable states in the mechanisms of gene replication. The instabilities are then corrected by the action of controlling elements, called *transposons*, that move along the genome to modify the expression of the coding sections by the removal or insertion of gene segments. Such modification results in chromosomal instability (leading to a loss of chromosomal segments during mitosis), or to the altered expression of certain genes (because of the insertion of new material into a sequence). The discovery of transposons, or *jumping genes* as they have been called, led to several interesting developments. It was demonstrated that the controlling system includes a regulatory element, acting as the *source* of a signal, and a target receptor gene, acting as the *sink*. Thus, the controlling system may be interpreted in terms of a *functional interaction*, as described above, the role of the vector being played by a diffusible, activating substance moving from the source to the sink. Although, here again we do not know whether the processes involved are strictly local, transposon dynamics may be reasonably considered to be of a non-local nature.

Finally, the discovery of transposons led to the development of gene dynamics, including studies of the spatiotemporal, topological variations of genes, not only in bacteria but also in higher organisms. The old notion of the permanence of heredity has

---

[25]As we shall see in Chapters IV and V, this property can be used to define the level of organization of the hierarchical system. By convention, the gene level will be identified with the zero-level in the hierarchy. The spatiotemporal variation of the products exchanged is significant during an interval of time that defines the time scale of the phenomenon occurring at this level.

gradually given way to the idea of a dynamic, adaptive evolutionary process based on the constant restructuring of DNA. *Thus, the genome may be considered in terms of the dynamic organization of a set of functional interactions.* The new findings in molecular genetics suggest that some of the statements, currently held to be established truths, such as the claim that each gene codes for a single protein, or that each operon regulates a specific physiological phenotype, should be considered with some caution since they do not constitute general rules. One of the more important consequences of the discovery of transposons is that these elements may explain the appearance of new characters in a species. Transposons, which combine readily with DNA, could facilitate the combination of genes of one species with those of another. Hereditary material would thus appear to possess the property of *plasticity*, suggesting a novel mechanism of biological evolution.

The functional organization of nervous tissue

For technical reasons, the brain became the object of a mechanistic approach only rather late in the history of science. At the beginning of the nineteenth century, Gall (1758–1828) introduced new methods of dissection that revealed some of the complexity of brain architecture, suggesting relationships between cerebral structures and functions. Unfortunately, Gall's misguided enthusiasm for phrenology, a pseudo-science according to which the shape of the skull allows an appreciation of mental qualities, brought him great discredit. Nevertheless, his anatomical techniques were put to good use by Bell (1774–1842) who published the first scientific outline of the anatomy of the brain, introducing the idea of nervous "circulation". In fact, Bell had discovered the phenomenon of double conduction — motor and sensory — in spinal nerves. Magendie (1783–1855) confirmed this finding and established precise relationships between the nerves of the body and the spinal cord.

As soon as Golgi (1844–1926) had developed his famous technique for staining nerve cells, Cajal (1852–1934) undertook a systematic morphological study of *neurons*, and the embry-onic development of nervous tissue. His painstaking research revealed that neurons are interconnected by means of special-ized anatomical structures called *synapses*. A typical neuron may be schematically considered as having three parts: the *soma*, or the cell body containing the nucleus; the *axon*, which is a long extension of the cell ending in an arborescence; and the *dendrites*, which are short branches of the cell. Synapses represents contacts between, on one hand, the terminal arborescences of the axon of one neuron (called the *presy-naptic* neuron) and, on the other, the dendrites or the soma of another neuron (called the *postsynaptic* neuron). The nervous system, which contains billions of neurons, each having hun-dreds of synapses, thus forms an unimaginably complex net-work. However, behind the apparently inextricable entanglement of axons and dendrites, there surely exists a wonderful organization just waiting to be explored.

Sherrington (1857–1952) showed that the essential functions of a neuron are the spatiotemporal *integration* of electrical sig-nals emanating from neurons that are connected to it, and the emission of an electrical potential, called the *action potential*, when a particular electrical signal — the *membrane potential* — exceeds a threshold value. Much research has been done on the *nervous influx* since the early work of Galvani (1737–1798) and Volta (1745–1827) on the electrical theory of muscle con-trol by nerves. Du Bois-Reymond (1818–1896) established the physical bases that were later brilliantly used in the description of electrical conduction and excitation in nerve fibers (Hodgkin and Huxley, 1952). These phenomena occur when sodium, potassium, chlorine and calcium ions cross the cell membrane, producing *ionic currents* that cause variations of the electrical potential inside the cell. The information trans-mitted appears to be essentially related to the electrical *activ-ity* of the cell, i.e. the frequency at which the action potentials are emitted by the neuron.

The synapse, which establishes a physical connection between two neurons, is generally considered the basic ele-ment in the organization of an elementary neuronal circuit. In

II

our representation of a physiological function, the synapse corresponds to a *structural discontinuity* in the neuronal medium. In physical systems, continuous structures are certainly capable of transmitting information over a distance, so why should there be a discontinuity in a physiological system? *The only reasonable explanation is that the discontinuity occurs at a level* different *from that of the neurons, implying the existence of a* new *physiological function.* This is an important aspect of the relationship between the structural hierarchy and the functioning of a physiological system at a given level. As we shall see in the case of the nervous system, the consideration of such discontinuities leads to an explanation of the functions of learning and memory.

The synapse has three important properties. First, it is almost punctual, with a size of only about a micron, constituting a very close link between two neurons; the *structure* of the synapse is localized, whereas — and this will be seen to be of importance — its *function* is not. Secondly, the synapse is *non-symmetrical,* being *oriented* from the presynaptic neuron towards the postsynaptic neuron. Finally, the *chemical* nature of the transmission leads to the release of a substance, called the *neurotransmitter,* which is capable of acting on the target cell.

The identification of modular brain structures and their functional properties would seem to offer a promising approach to the investigation of cognitive functions. Some attempts have been made along these lines, taking into account the hierarchy of the structures involved. For example, certain *macromolecules,* which intervene in synaptic reactions, belong to the *membrane channels* situated inside the *synapse.* The connections and interactions between these components are called *microcircuits* (Shepherd, 1992). Such microcircuits can be regrouped as *dendritic subunits* within the *dendritic tree* of a neuron. Then, continuing from the dendritic tree and working our way up the structural hierarchy, we come to the *neuron,* and then to the *local circuits* made up of interacting neurons and, beyond these, to *groups of interconnected neurons* related to different parts of the brain. Here we have *eight levels of structural hierarchy.* However, it has to be admitted that the choice of these levels and the identification of the corresponding functions are largely arbitrary. This is why it would be preferable to reason in terms of the identification of functional interactions and on the stability of the associated physiological systems. This approach has been successfully used in the case of the cerebellar cortex (Chauvet and Chauvet, 1993).

Let us now consider some results obtained by electrophysiological methods. The results of experiments on neurons and synapses are interpreted in terms of two electrical quantities, i.e. the *potential* and the *conductivity*. From the functional point of view, the presynaptic neuron emits an action potential at the *axon hillock*, i.e. the region in which the axon originates from the neuronal cell body. This is the *source*, from which the action potential is actively propagated, with no loss of amplitude, along the axon towards its extremity. At the extremity of the axon, the action potential triggers the release of the neurotransmitter that, after diffusion across the synaptic cleft, binds to a receptor on the postsynaptic membrane, modifying its electrical potential. The receptor is thus the *sink*, located at the lower level of the synapses in the postsynaptic neuron. The chemical phenomena that occur involve membrane structures situated at a distance and intracellular organelles. The integration of thousands of such postsynaptic membrane "micropotentials" then leads to a new value of the membrane potential at the axon hillock of the postsynaptic neuron. Thus, in the global sense, a membrane potential is transported from one axon hillock to another by means of a certain number of transformations within the neuron. The membrane potential, which may be considered as the vector of the *functional interaction*, is propagated from one point to another, situated at a certain distance, in this new space of neuronal axon hillocks (Figure II.5). Here again, we may observe that the phenomena described are not strictly local.

Let us imagine a "selective" microscope that would allow us to choose the level of structural organization observed in a sample of nervous tissue, each level of organization corresponding to a certain "magnification". Figure II.6 illustrates some of the views that could be obtained with such a microscope. The

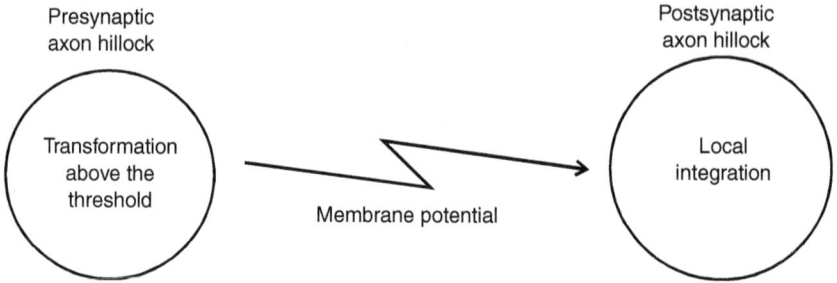

Figure II.5: The non-local functional interaction between two neurons corresponds to the membrane potential.

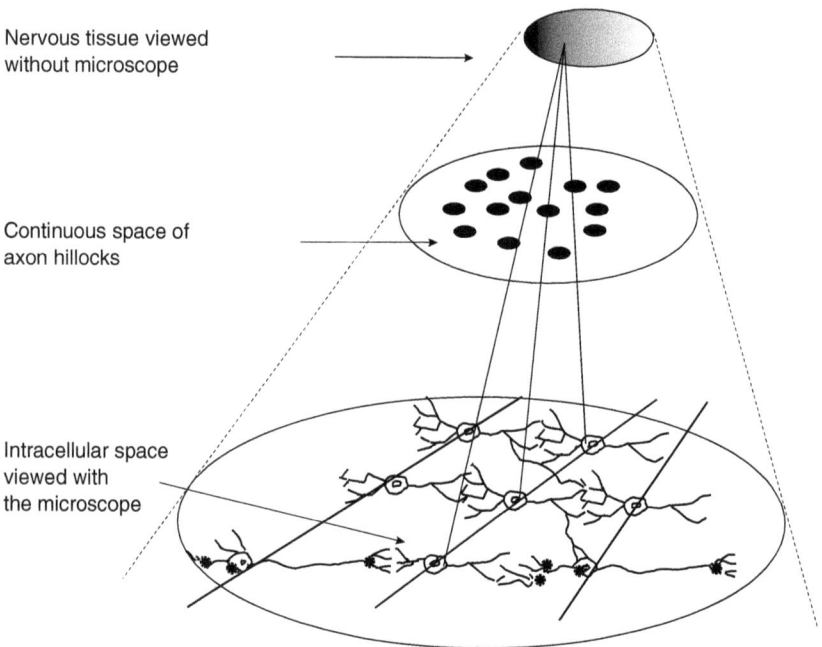

Figure II.6: Nervous tissue seen through an imaginary "selective" microscope. The unmagnified image represents nervous tissue at the highest level of structural organization, at which none of the individual structures are visible, whereas the magnified images correspond to two "selected" levels of structural organization, i.e. the space of axon hillocks, and the space of synapses.

unmagnified image would correspond to the highest level of structural organization, represented by a continuous space with no visible structures. The highest magnification would correspond to the lowest level of structural organization, showing all the structures of nervous tissue. We might expect an intermediate magnification to reveal an intermediate level of structural organization, e.g. the space of axon hillocks. Of course, the 'selectivity' of the microscope should be such that, in this view, only the axon hillocks will be visible, all the other parts of neurons remaining invisible.

In Figure II.6, straight lines from the axon hillocks at the lowest level of structural organization may be projected through the space of axon hillocks at the intermediate level right up to the highest level. We may then visualize other structures playing a role in the emission of an action potential, for example, the membrane channels that open when a neurotransmitter binds to a membrane receptor. This would of course reveal a new level of structural organization. Thus, we see how the structural hierarchy of a system is generated, and how it is related to the property of non-locality. As in the case of the biochemical and epigenetic networks considered above, the organization of nervous tissue shows that *structural discontinuity* is at the origin of non-locality. Here, the set of functional interactions, which diverge from the neuron-source to the set of neuron-sinks, may be represented by the propagation of the membrane potential from one axon hillock to another in the neuronal network. In effect, all the neurons of a given network cooperate to produce a certain activity, for instance, the learning and the memory of a task. The collective behavior of the set of functional units defines a physiological function. However, the physical representation of this function is greatly complicated by the fact that the structural hierarchy, which implies non-locality, is associated with a *functional hierarchy.*

Again, by adjusting the "magnification" of our imaginary microscope so as to visualize the appropriate structural level of nervous tissue, we would see the space of synapses. Here we would also see other intracellular structures, such as various elements of the cytoskeleton, acting as structural discontinuities. In other words, we would be observing highly complex phenomena since there is much more going on at the synaptic connections than the mere diffusion of neurotransmitters according to the laws of diffusion in aqueous solutions, or the transport across membranes by specific transporter molecules. These connections actually modify the efficacy of synaptic transmission on a much longer time scale (ranging from about a second to several minutes) than that of the observed changes in the membrane potential (measured in milliseconds), thus modulating nervous activity in the corresponding neuronal network. This modulation can be measured by a quantity called the *synaptic efficacy*, i.e. the strength of a synaptic connection between the presynaptic and the postsynaptic elements, the value of which at a given instant depends on the molecular processes involved in the opening and closing of the membrane channels in the synapses.

The analysis of experimental observations leads us to certain general concepts of structural discontinuity and hierarchy. Just as the membrane potential represents the functional interaction in the space of axon hillocks at the level of the neuronal network, the synaptic efficacy in the space of synapses corresponds to the functional interaction at the level of the synaptic network. Since the structural discontinuities are perfectly identifiable in these cases, the non-local functional interaction can be determined. It is true that the choice of the variable measured is somewhat arbitrary, but only within certain limits since it must of course be observable. As we shall see in Chapter VI, the choice of the variable may also be made taking into consideration the field theory we propose. For the time being, we may say that one structure (synapse A) acts on another (synapse B) through a functional interaction representing the modification of the number and conductance of open membrane channels. The formal schema in Figure II.1 may again be used to represent this interaction (Figure II.7). The global behavior of the set of synapses, taken together with the extra-synaptic elements, produces the physiological function corresponding to synaptic modification that we may call the "heterosynaptic efficacy". As

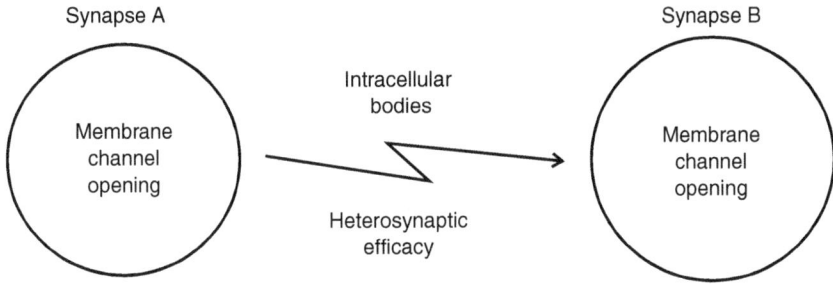

Figure II.7: The functional interaction between two synapses corresponds to a variation of synaptic efficacy from one synapse to another.

mentioned above, this physiological function operates on a specific time scale of the order of seconds to minutes.

Similarly, we may extend this generalization to higher levels of organization under the condition that certain rules be satisfied. For example, the association of groups of neurons, with a functional interaction, which may be called "*group activity*", between one group of neurons and another, corresponds to a higher level of organization than that considered above. This is the case in brain structures such as the nuclei of the brain stem, or the cerebellum, or again the hippocampus. We may then go on to consider larger structures, made up of groups of neurons, and organized according to the functional interactions between these groups. In principle, there is no difference between this example and those discussed above, such as the interaction between an organ, e.g. the pancreas, and certain cells of the body, e.g. the hepatocytes.

Let us now consider the classic example of the *substantia nigra*, a cerebral nucleus known to be involved in Parkinson's disease. The dysfunction of the substantia nigra causes the disease when the amount of *dopamine*, a major neurotransmitter, falls below a certain level. In the normal state, the substantia nigra and another cerebral nucleus, called the *corpus striatum*, both have high levels of dopamine. After it is produced in the substantia nigra, the dopamine migrates along nerve fibers to act at a distance, inhibiting neuronal activity in the corpus striatum. The feedback loop, operating from the corpus striatum to the substantia nigra, has an entirely different type of action since it is not dopaminergic but GABAergic. The dopaminergic system plays an important role in the control of involuntary movements, such as the automatic movements required to maintain the body's balance. It is therefore easy to see why a perturbation of the function of the substantia nigra should lead to Parkinson's disease. In this formulation, the control of the involuntary movement of the body is ensured by an element of functional organization in which the feedback between the substantia nigra and the corpus striatum is a *non-local, non-symmetrical functional interaction*.

Structural hierarchy and functional hierarchy in living organisms

The analysis of several biological phenomena has led to the identification of *functional interactions* occurring between the *structures* of a living organism. We shall refer to these structures as *structural units* to stress the underlying reality of the abstract organization of these units. As we shall see in the following chapters, *the structural units are in fact material, identifiable elements that ensure the dynamic stability of the corresponding physiological system.*

What conclusions can we draw from the examples considered above?

The functional interaction between one structure and another must occur across one or more structural discontinuities, since the source and the sink are necessarily situated at a certain distance from each other in physical space. Functional interactions may occur, for example, between one local chemical reaction and another (Figure II.1), between a pancreatic cell and some other cell in the body (Figure II.2), between the chromatin and the cytoplasm (Figure II.4), between a presynaptic neuron and a postsynaptic neuron (Figure II.5), or between one synapse and another (Figure II.7). The structural discontinuities lead to the property of the non-locality of the functional interaction.

The structural discontinuities may be true physical structures, e.g. cell membranes, or limits across which physical properties in contiguous tissues or systems differ, e.g. the cardiovascular system in Figure II.2.

The unity and the individuality of a living organism imply the global regulation of its functions through homeostasis. As we shall see in the next chapter, homeostasis ensures the internal equilibrium of the body. Since living processes comprise a set of closely regulated local transformations, which are in fact chemical reactions at the molecular level, interactions must occur between structures that are necessarily distant from one another in physical space. This corresponds to the property of non-locality. If structural discontinuities are present, a hierarchy of functional interactions will emerge. Thus, the property of non-locality is associated with that of a functional hierarchy. To take another example, the process of *cell specialization* occurs in distant parts of the body. This is important from the functional point of view since the process imposes a functional interaction by means of which the information associated with cell specialization is transmitted to other structures of the body, again implying the property of non-locality.

In the course of this chapter, we have considered several important concepts, such as those of functional interactions, structural discontinuities, non-locality, structural hierarchy, functional hierarchy and cell specialization. It would appear difficult to establish a classification of these concepts in terms of causality. It may be admitted that structural discontinuities lead to non-locality, probably because of the specialization involved. However, does the structural hierarchy produce the functional hierarchy, or is the opposite true? We may even wonder whether this question really makes sense. In any case, it would appear more judicious to consider the problem in terms of the coupling between the geometry of the system, as represented by its structural organization, and its topology, as represented by its functional organization. This is what we propose to do in the following chapters.

III

THE INTEGRATION OF PHYSIOLOGICAL FUNCTIONS

Homeostasis in Living Organisms

III

THE INTEGRATION OF PHYSIOLOGICAL FUNCTIONS

Homeostasis in Living Organisms

The conundrum of life has challenged thinkers all through the ages but no satisfactory solution has ever been found. It is true that any attempt to explain the integration of physiological functions in living organisms is somewhat like defining life itself. However, our aim here is rather more modest. We propose merely to formulate certain questions concerning physiological functions, and examine some of the consequences. As described in Chapter I, physicists analyze a given physical system, for instance in terms of matter and energy, by determining the *invariants* of the physical system, the knowledge of which is essential to the understanding of the changes occurring in the system. We propose to use a similar theoretical approach for biological systems. Biologists have long been intrigued by the *homeostasis*, or invariance, of living organisms in a surprising variety of conditions. Claude Bernard (1865) was the first to draw attention to the regulation of the internal medium of an organism with respect to the fluctuations of the external environment:

> "The fixity of the internal medium is the main condition of a free, independent life; the mechanism ensuring this involves the maintenance of all the conditions necessary to the life of the elements contained therein."

III

Anticipating the way of thinking of the modern cyberneticists, Claude Bernard reasoned that the normal, uninterrupted functioning of an organism depends on mechanisms of internal regulation capable of reacting to environmental changes. The organism is thus viewed as a self-controlled system, with a degree of efficiency matching its evolution. Of course, this mode of operation is not specific to living organisms and any engineer today knows how to construct mechanical systems based on the principle of dynamic regulation.

In Chapter II we discussed the cybernetic nature of enzymatic regulation at the genetic level. We may recall that the operation of the system involves the activation of *structural genes* (coding genes) by *operator genes* under the control of *regulator genes.* The regulator genes allow the synthesis of repressor proteins that switch the operator genes on or off according to the amount of enzymes synthesized by the structural genes. The behaviour of this type of system has been mathematically described (Babloyantz and Nicolis, 1972). In fact, there are many other examples of cybernetic regulation at the cellular level. We may mention the modification of membrane permeability that allows the coordination of the different metabolic activities of the cell; the retro-inhibition of a metabolic pathway by the action of the final product on the first step; or again, the law of mass action that regulates the equilibrium concentrations in chemical reactions. All these mechanisms are based on the phenomenon of *homeostasis,* an essential feature in biological systems. Homeostasis maintains the biological system in a stationary state, i.e. a *state of dynamic equilibrium,* which differs from the state of thermodynamic equilibrium. In a system that is essentially open, the fluxes of matter and energy are maintained constant by the dynamic equilibrium.

The etymology of the word "homeostasis" from the Greek — *homoios* (equal) and *stasis* (state) — suggests that the term refers to the stabilization of a state. Claude Bernard's ideas on the fixity of the internal medium appear to prefigure the concept of biological homeostasis. W. B. Cannon first used the word in the modern sense in his celebrated article "Organization for physiological homeostasis." (*Physiological Review*, 9, 399–431, 1929). It is worth citing his original definition:

> "Higher living organisms constitute an open system with many relationships with the environment. Changes in the environment affect the system directly or cause it to react, thus leading to perturbations in the system. Such perturbations are normally kept within close limits because automatic adjustments take place inside the system so as to avoid large oscillations while the internal conditions are maintained more or less constant [...]. The coordinated physiological reactions that maintain most of the dynamic equilibria of living organisms are so complex and so specific that these should be designated by the special term of homeostasis." (W. B. Cannon, *The Wisdom of the Body*, New York, 1939.)

It was only much later that Prigogine *et al.* (1972) investigated dynamic physical equilibria using thermodynamics. The thermodynamics of non-equilibrium states has opened up many new lines of research in physics as well as biology. As we shall see in this and the following chapters, there are fundamental differences between the thermodynamics of the non-equilibrium states of physical structures and those of physiological functions. However, the common point is the mode of representation, which is based on the mathematical theory of dynamic systems, and the study of these systems in the neighborhood of stationary states.

Thus, if the survival of an organism requires that it be maintained in equilibrium with respect to the medium in terms of energy, then it must acquire the ability to function in an optimum manner. This statement could be challenged for its finalism but for the observation of the abundance of cybernetic regulations at work in elementary physiological mechanisms. How are the invariants mentioned above maintained during the development of an organism and its evolution? How is the composition of the internal medium kept constant? All the functions of a living organism are so closely linked that the disorder of any one of the functions is bound to affect all the others. A satisfactory theory of functional organization will therefore have to explain the global regulation of the organism.

III

Strictly speaking, in terms of self-organization, the notion of homeostasis is not applicable to physical systems. Such systems have no history, in the sense that they do not evolve through self-reproduction. In other words, physical systems cannot accumulate the selective advantages of regulatory loops from generation to generation. The concept of homeostasis was introduced to describe the complex functional relationships that ensure the stability of living organisms. It is obvious that in a biological system, inasmuch as it is made up of matter, the laws of physics must apply. In particular, since it is an open system, the principles of thermodynamics must hold as well. However, it is equally obvious that the biological system, considered from a functional point of view, obeys specific laws governing the regulatory mechanisms that operate at different structural levels of the organism to maintain the homeostasis of the internal medium. Homeostasis is thus the end-result of the harmonious integration of physiological functions. Should some event intervene to upset this delicate balance, disease is likely to appear. However, the official recognition of this fact did not come easily. In the first half of the twentieth century, some physicians had already identified certain clinical syndromes connected with greatly disturbed hydroelectrolytic equilibrium. Such was the case, for example, in cholera patients who suffered heavy losses of potassium during bouts of diarrhea. However, in the absence of sound biological analysis, these clinical observations were long ignored. In the early 1960s, clinical explorations and fundamental research revealed the role played by osmosis in the equilibrium of water. Since then, the mechanisms of regulation and excretion of water and electrolytes have been linked with hormonal and renal activity.

Claude Bernard's concept of the internal medium

Considering that 60–70% of the total body weight of living organisms consists of water, Claude Bernard (1865) wrote:

> "No one will contest the importance of the study of the different liquids of the body in the normal and the pathological states. It is in the blood and in the liquids deriving from blood

that physiology finds most of the conditions required for the accomplishment of the physicochemical acts of life, and it is in the alterations of these same liquids that medicine seeks the causes of a great number of diseases."

An examination of body liquids gives a good picture of the internal dynamic equilibrium of the organism. Water in the body is distributed between two major compartments — intracellular and extracellular. Claude Bernard called the extra-cellular compartment, in which the cells bathe and live, the internal medium. This compartment contains blood and lymph as well as the various substances produced, for example, by the digestive or urinary systems (Fig. III.1). Some scientists believe that the internal medium actually reflects the composi-tion of the sea in prehistoric times when our Cambrian pre-cursors may have found it advantageous to include a little of the sea into their teguments, thus making it a part of them-selves. Blood contains, on one hand, plasma with its macro-molecules, proteins, globulins and ionized as well as non-ionized substances, and, on the other, formed elements, mainly red blood cells and antibodies. Thus, the internal medium transports indispensable nutriments to all parts of the body and carries away waste products so that the cells resist aggression and remain alive. The extracellular liquids, blood plasma and lymph, contain the ions and electrolytes that ensure a dynamic equilibrium between all the compartments of the body.[26] But what exactly are the mechanisms involved in maintaining the dynamic equilibrium called *homeostasis?*

Let us briefly examine the relationships between the major physiological functions of the body. It is interesting to analyze the physical basis of the exchanges that occur and the organs that are involved. Thermodynamics can determine the pressure exercised by a liquid in terms of its concentration in particles, molecules or ions. Let us consider two compartments separated by a membrane. The osmotic pressure is defined as

---

[26]This means that the concentration of molecules in the internal medium, after perturbation, tends to a stable equilibrium point of the dynamic system describ-ing the process. For more details, see the section on stability in Chapter IV.

III

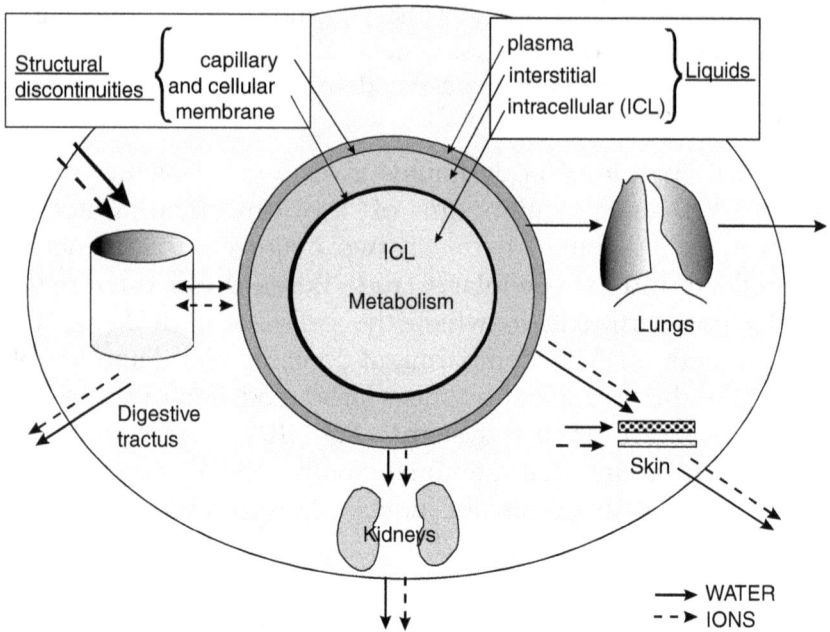

Figure III.1: The fluxes of water and ions between the digestive organs, the lungs, the skin, the kidneys and the intracellular liquids (ICL), the extra-cellular liquids (Claude Bernard's internal medium), the interstitial liquids and the secretions. The structural discontinuities, here represented by capillary membranes and cell membranes, are at the origin of the non-local effects observed.

the difference in pressure between the two compartments containing solutions not having the same molecular concentrations of the solute.[27] We may recall that the pressure acting against a wall is due to the Brownian movement of particles striking against the wall. Since the number of particles is not the same on both sides of the membrane, a difference of pressure arises and, if the membrane is permeable, water will be transported across the membrane until thermodynamic equilibrium is attained. In fact, the exchanges between the two compartments

---

[27]The osmotic pressure of a solution measures the decrease in concentration produced by the solute. This is expressed by $\Delta\Pi = P_A - P_B = RT(c^A - c^B)$. P_A and P_B are the pressures in the compartments A and B separated by a semi-permeable membrane; c^A and c^B are the concentrations of the solute in the two compartments; R is the perfect gas constant; and T is the temperature.

are initiated by the difference of pressure between the compartments. The quantity of molecules and ions present in each of the compartments, independently of the quantity of water, plays a fundamental role in determining the osmotic pressure. Thus, according to the osmotic pressure acting on a cell, it will tend to either fill up with water or empty itself.

Let us now consider what happens when the heart pumps blood through the capillaries, which are tiny tubes with a diameter of the same order as the red blood cells. Blood contains large amounts of various proteins that play the role of the solute. Since the plasma proteins cannot cross the capillary membrane, the osmotic effect produced has to be balanced by the tendency of water and solutes to migrate from the interstitial space to the plasma space. The hydrostatic forces produced are essential to the exchanges occurring at the capillary level. However, there is another hydrostatic force simultaneously at work, due to the blood being pumped by the heart, which tends to expel water and solutes from the capillaries. These two phenomena — the hydrostatic forces on either side of the capillary membrane and the circulatory hydrostatic pressure — are antagonistic. Vascular pressure plays a major role and the transfer of water results from the interplay between blood protein content and blood pressure. Clearly, any disorder in one medium will immediately affect the other. From the medical point of view, this constitutes an advantage since it is easy to monitor the composition of the extracellular liquid by analyzing blood plasma. In some extreme situations, body tissues may be infiltrated by water displaced from the extracellular to the intracellular compartment, causing edema. This condition is observed, for example, in cases of protein deficiency in the blood or severe dehydration following third-degree burns. In short, the transport of water may be described in terms of the rupture of a functional dynamic equilibrium in a system seeking a physical thermodynamic equilibrium.

Homeostasis and functional equilibria

In its normal condition, the body is not in a state of thermodynamic equilibrium. In fact, true thermodynamic equilibrium would correspond to death. So how do living organisms keep away from this fatal equilibrium state? The simple answer is that in living creatures an uptake of energy ensures the unequal distribution of ions.[28] In fact, the combined action of three mechanisms — cell metabolism, osmotic forces and hydrostatic forces — produces a network of stationary fluxes, principally of water and sodium, in the whole body. Specific organs regulate these fluxes: the digestive system controls the input and output of organic and inorganic matter, while the lungs, the skin and, in particular, the kidneys control the circulation of water and the electrolytes. Liquids flow through all parts of the body carrying raw material for processing, removing waste matter and transporting finished products to their destination.

What happens in the body is reminiscent of industrial activity in an organized human society. This analogy is worth exploring. In both cases, the systems depend on organized structures. Yet, no one — not even the most fervent adept of orthodox physics — would claim that the organization of human society is merely of a structural nature. As in the case of human society, the organization of an individual organism is essentially functional. Naturally, this does not mean that the organism escapes the laws of physics. For example, a hormone binds to its receptors according to the law of mass action just as an object remains on a table under the law of gravity. However, the hormone itself is emitted by an endocrine gland according to the laws of functional biological organization in a manner similar to the production of consumer goods by a factory operating under the laws of market forces. In both these

---

[28]The distribution of ions in an organism varies according to the sectors considered. In the interstitial sector, the protein concentration is much lower than in the intravascular and cellular compartments; in the extracellular sector, the concentration of sodium and chlorine ions is much higher than in the intracellular sector; in the intracellular compartment, the concentration of potassium is much higher than in the extracellular compartment. This unequal distribution of ions is due to the mode of transport of sodium, i.e. active transport, across the cell membrane. The energy required for the active transport is extracted in the form of ATP in the course of cell metabolism as described in Chapter II.

situations, the notion of homeostasis is applicable. In the industrial context, homeostasis would concern the functional economic organization involved and would be identified with the maintenance of a dynamic equilibrium of the society. In contrast, in a living organism, homeostasis is the permanent condition of existence. Of course, the time scales that apply to individuals are very different from those that apply to societies.

Fundamentally, the unity of an organism is evidenced by the integration of its physiological functions through the stabilization of the internal medium, the sole object of which is to ensure an adequate capillary flux of suitably mineralized water. The homeostasis of the "internal sea" corresponds to the permanent dynamic equilibrium between the liquid-containing compartments. This is maintained by three physico-chemical mechanisms: the thermodynamic equilibrium of water and electrolytes between the cells and the interstitial liquids ensured by *cell metabolism* and *osmotic forces*; the exchanges between the interstitial liquids and plasma produced by *osmotic forces*; and the action of *hydrostatic forces*. The exchanges occur through membranes with selective permeabilities. In fact, these membranes constitute structural discontinuities in the biological system. Mechanisms operating *inside* the structural discontinuities regulate the composition of the internal medium. This description of homeostasis clearly implies the existence of a structural hierarchy.

All the physiological functions of a living organism contribute to its homeostasis. The digestive function provides the energy required for active exchanges; the respiratory function supplies oxygen and, together with the renal function, removes the hydrogen ions resulting from metabolic activity; the circulatory function maintains a constant hydrostatic pressure in the capillaries; and the renal function regulates the ion concentration in the body liquids. These physiological functions are controlled by the nervous and endocrine systems. All the partial regulations involved can be integrated, as we shall see below, into two major functions: the *hydroelectrolytic equilibrium*, regulating water and sodium; and the *acid-base equilibrium*, regulating carbon dioxide.

III

The problem that arises is to determine the relationship between the several physiological functions involved in the homeostasis of an organism and to establish the links between the functional organization and the underlying structural organization. We propose a solution based on the construction of a functional hierarchy linked to a structural hierarchy. In this construction, the integrated functional organization of the body is represented by a set of diagrams showing the points of action of the various physiological functions of the organism. Some interesting models have been suggested for the formalization of this complex set of relationships at the highest level of organization in the body. For example, Grödins *et al.* (1967) studied respiratory regulation, taking into account the perturbations of the acid-base equilibrium; Guyton *et al.* (1972) described cardiovascular regulation, including several physiological functions; and Koushanpour and Stipp (1982) investigated the regulation of body liquids. However, all these models suffer from over-simplification because of lack of information concerning several of the parameters involved.

Since the internal medium of an organism is in a state of dynamic equilibrium, it would seem useful to investigate the major systems of the body in terms of the stability of their organization rather than through analytic studies. We therefore propose to describe the functional organization of a biological organism, stressing the notion of structural discontinuity, to bring out the relationships between the different systems of regulation. First, let us illustrate the integrated functional organization of an organism by means of the two main equilibria of the body — the hydroelectrolytic and acid-base equilibria.

The essential environmental functions of the integrated functional organization

The most striking evidence of the existence of an integrated functional organization in a living organism is found in the equilibrium of ions, electrolytes and water, established while a patient undergoes reanimation. This integration can literally be "seen", minute by minute, as the monitoring systems churn out data revealing the ongoing changes in the biological parameters of the patient, reflecting the course of the treatment administered. But how are we to understand, i.e. literally "integrate" into our thinking process, the huge mass of figures representing, for instance, the continuously varying concentrations of ions and molecules in the blood, corresponding to the events occurring within the body, even deep inside the cells, under the action of complex physical and physiological mechanisms? In fact, what we have here is the image of the dynamic functional organization of the various organs of the body, controlled by the nervous and endocrine systems, protected by the defense system, all aimed at maintaining a delicate physiological equilibrium. Surely it is most extraordinary that a "machine" as complicated as a living being can continue to operate even in severely hostile environments, constantly subjected to internal disorders[29] as well as the external constraints of temperature and availability of food.

Let us recall the mechanism of cellular respiration, described in Chapter II as being "*silent*" in contrast to the "*noisy*" respiration of the lungs. Cellular respiration is regulated by the demand for ATP, i.e. according to the energy needs of the cell. This metabolism leads to the formation of several acid byproducts so that the concentration of hydrogen ions needs to be regulated. The law of mass action applies to the carbon dioxide and water, carbonic acid, hydrogen ions and bicarbonates.[30] The lungs eliminate the volatile carbon dioxide while the bicarbonates[31] are recycled in the kidneys where, after filtration, they

---

[29]For example, the derepression of cell division, which may occur at any moment in a cell, is believed to lead to the permanent development of multiple micro-cancers in the body.

[30]These five molecules interact according to the equilibrium equation: $CO_2 + H_2O \leftrightarrow H_2CO_3 \leftrightarrow H^+ + HCO_3^-$.

[31]The bicarbonate-carbon dioxide couple (HCO_3^-/CO_2) is an excellent buffer system — the best in the body. It is also the most important because of the volatility of CO_2. However, this raises several problems: (i) the base of the buffer system: $HA \leftrightarrow H^+ + A^-$ has to be regenerated; (ii) the non-volatile acids such as the sulfuric and phosphoric acids, lead to the depletion of plasma bicarbonate.

are finally absorbed. This chemical decomposition of carbonic acid makes the pulmonary and renal functions interdependent. Moreover, the excretion of hydrogen and potassium ions, and the reabsorption of sodium ions, bicarbonates and water are closely linked, so that there is a strong relationship between the acid-base and hydroelectrolytic regulations.

Hydroelectrolytic homeostasis ensures the regulation of water and electrolytes such as sodium, potassium, chlorine, and so on. This regulation, also known as the "regulation of volume" since large amounts of water and electrolytes are displaced in the organism, involves three fundamental factors: cell metabolism, osmotic pressure and hydrostatic pressure. The fluxes of water and ions are regulated at the capillary level all through the body, and in the kidneys. The controls, operating at a distance, are essentially hormonal. The endocrine organs contributing to the regulation are the post-hypophysis located in the brain, the cortico-surrenal glands and the kidneys. These organs secrete, respectively, the antidiuretic hormone (ADH) that allows the kidneys to retain water, aldosterone that controls the excretion of salt, and renin that modulates the caliber of the blood vessels in conjunction with angiotensin, under the control of the autonomous nervous system.

The renin-angiotensin-aldosterone (RAA) system is controlled by the autonomous nervous system as well as the cardiovascular system. The arterial pressure acts on the intrarenal and systemic baroreceptors. The former operate directly

---

(*Footnote 31 Continued*)

In general, the quantity involved s very small (50–100 nmol H^+/day), but in certain pathological conditions the decrease in the concentration of HCO_3^- leads to an increase in CO_2; and (iii) inversely, the excessive excretion of H^+ ions leads to a basic behavior, equivalent to the excessive retention of a base.

The acidity of a medium is measured by its pH. The equation: $Acid \leftrightarrow H^+ + Base$ defines the equilibrium constant: $K = [H^+][Base]/[Acid]$, which may be written in logarithmic form to give: $pH = -\log [H^+] = -\log K + \log[Base]/[Acid] = pK + \log[Base]/[Acid]$. This is known as the *Henderson-Hasselbach equation*. For the buffer HCO_3^-/CO_2, which results from the two equilibrium reactions, we generally take into account the global equilibrium of the dissociation constant K' such that: $pK' = -\log ([H+][HCO_3^-])/([CO_2]+[H_2CO_3])$.

whereas the latter increase the secretion of renin through orthosympathetic nervous activity leading to the release of circulating catecholamines. A change in arterial pressure acts directly on glomerular filtration and renal blood flow, modifying the reabsorption of water and salt. In the RAA system, angiotensin II, resulting from the transformation of angiotensinogen into angiotensin I under the action of renin secreted by kidney cells, stimulates the release of aldosterone. This constitutes the first feedback loop. The second feedback loop operates directly on the arterial pressure through the reduction of capillary diameter under the action of angiotensin I. The third feedback loop includes the hypothalamic osmoreceptors that control the release of ADH, which produces a sensation of thirst in the organism. Thus, an increase in the secretion of renin leads to the following events: local self-regulation of the secretion of angiotensin II, increase in arterial pressure, increase in the secretion of aldosterone, increase in the reabsorption of sodium, increase in the secretion of ADH, and increase in the reabsorption of water. The final result is the regulation of the extracellular volume (Fig. III.2).

The complex cybernetic nature of the RAA system is beautifully illustrated by the operation of various organs at a distance from the site of action and the several interlocking feedback loops that are brought into play. The intricacy of the mechanisms involved makes the representation and analysis of the system particularly difficult. The RAA system, which is of considerable medical interest, exercises a four-fold action in the organism: regulation of arterial pressure; regulation of the blood mass; homeostasis of ionic pressure (hydrogen, oxygen and carbon dioxide); and homeostasis of metabolism, in particular the metabolism of sugars.

How the body gets rid of excess acidity produced by the metabolism

The brain, the noble organ of the body, poses a major environmental problem in the organism. It consumes large amounts of glucose and burns up enormous quantities of

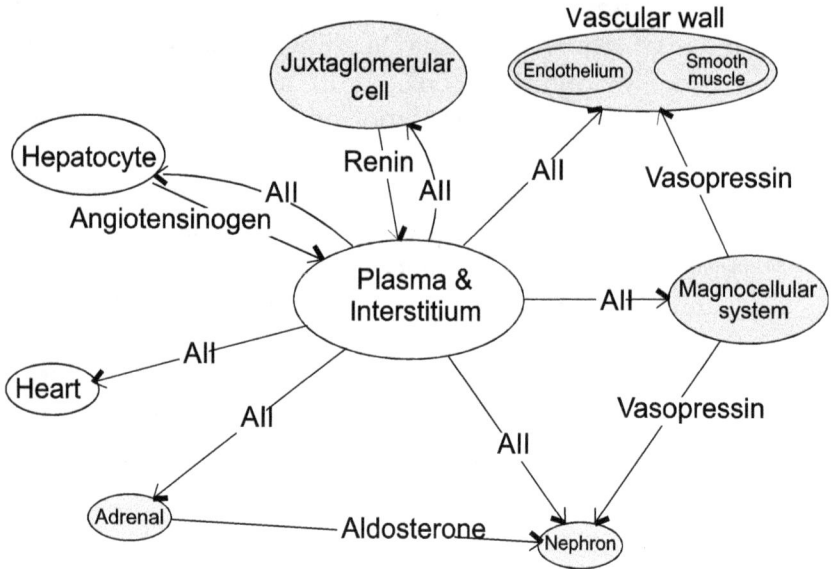

Figure III.2: The RAA system showing the functional interactions between the different organs. The interactions operate *via* the body fluids, plasma and the interstitium. *All* represents the set of molecules.

oxygen (25% of the total body consumption whereas it represents only about 2% of the mass), producing a high level of acidity. The brain is as well-protected against mechanical shocks by the skull and its meningeal structures as it is against bacterial attack by the blood-brain barrier. But it also needs to be protected against the excess acidity produced by the metabolic activity of cerebral tissue. The by-products of metabolism, hydrogen ions and carbon dioxide, diffuse into the cerebrospinal fluid (CSF) that may be truly qualified as the "internal sea". The CSF, a medium that has still to reveal all its secrets, contains neuronal endings that act as the chemoreceptors of a highly efficient regulatory system designed to eliminate wastes (Fig. III.3).

Some knowledge of anatomy is necessary to the understanding of the working of the neurons involved in the regulation of the CSF. These neurons, situated in the medullar region at the upper end of the spinal cord, bathe in the interstitial cerebral fluid. Although separated from the CSF, this

RN

X

CB

AB

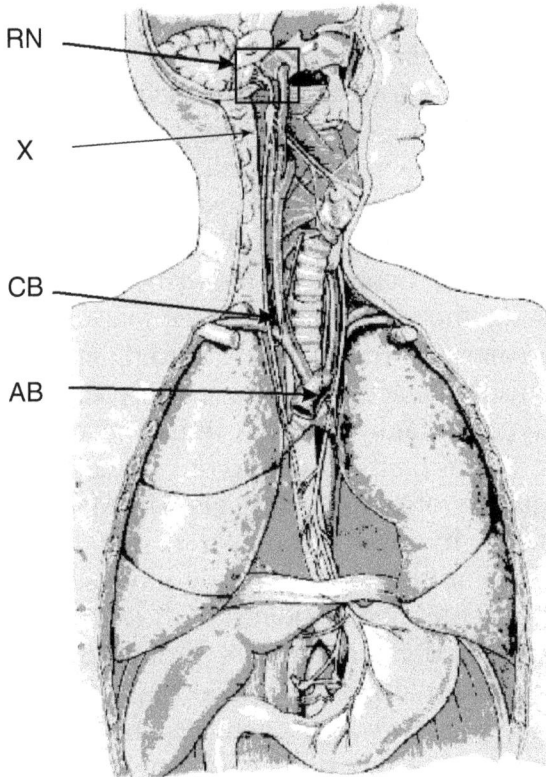

Figure III.3: The respiratory neurons (RN) situated in the medullar region (square) of the brain carry out the central regulation of the respiratory function. This illustration shows the position of the peripheral chemoreceptors: the carotid body (CB), the aortic body (AB), and the pneumogastric, vagal or X nerve. The left and right branches of the X nerve innervate the digestive organs and the cardiovascular system.

group of neurons, being highly permeable to electrolytes and carbon dioxide, may be considered as forming a single compartment. In fact, the neuronal receptors sensitive to the variation of the concentration of these substances are situated precisely in the interstitial cerebral fluid. The surrounding compartments are vascular compartments (capillaries) that are almost impermeable to large molecules and ions, particularly the bicarbonate and hydrogen ions. Thus, the regulation of the acid-base equilibrium requires energy for the active transport of molecules and ions. This is a slow chemical process, with a reaction time of 20–40 sec.

In contrast, the peripheral arterial chemoreceptors stimulate respiratory centers in less than 1 sec. The chemoreceptors are made up of small groups of specialized cells, constituting the aortic and carotid bodies, situated against the walls of the aorta and the carotid artery. These bodies contain chemosensitive cells surrounded by neurons of the sympathetic nervous system, suggesting the existence of a feedback loop providing highly precise nervous regulation. According to Jacquez (1979), the intense metabolic activity of these richly vascularized structures maintains the high detection threshold of the receptors. The increase in the partial pressure of O_2 and the pH leads to a higher frequency of action potentials. The afferent nerve endings respond to hypoxia through a neurotransmitter that is released by the carotid bodies into the bloodstream.

Finally, the efferent nerve endings of the neural system act on the thoracic and abdominal muscles through the motoneurons. A specialized neuronal network periodically stimulates groups of respiratory muscles, *via* the pneumogastric, vagal or X nerve (Fig. III.3), to modify the ventilatory activity. Higher nervous centers receiving information concerning this activity then determine the degree of ventilation required by the body. This corresponds to a mechanism of neuromechanical regulation.

Figure III.4 schematizes the regulation of the chemical, neurochemical and neuromechanical components of "silent" cellular respiration in terms of functional interactions. This highly abstract, conceptual representation of the phenomenon contrasts sharply with the classic illustration of the anatomical structures involved in respiratory activity, as shown in Fig. III.3.

What happens to the metabolic waste products after they enter the blood stream? Aerobic metabolism imposes the same constraints on all animal organisms: the absorption of oxygen, the rejection of carbon dioxide produced by the oxidative cellular metabolism, and the regulation of the equilibrium of water. The essential function of the metabolism is to bring oxygen to the cells, tissues and organs of the body and to remove carbon dioxide. In human beings, the respiratory system

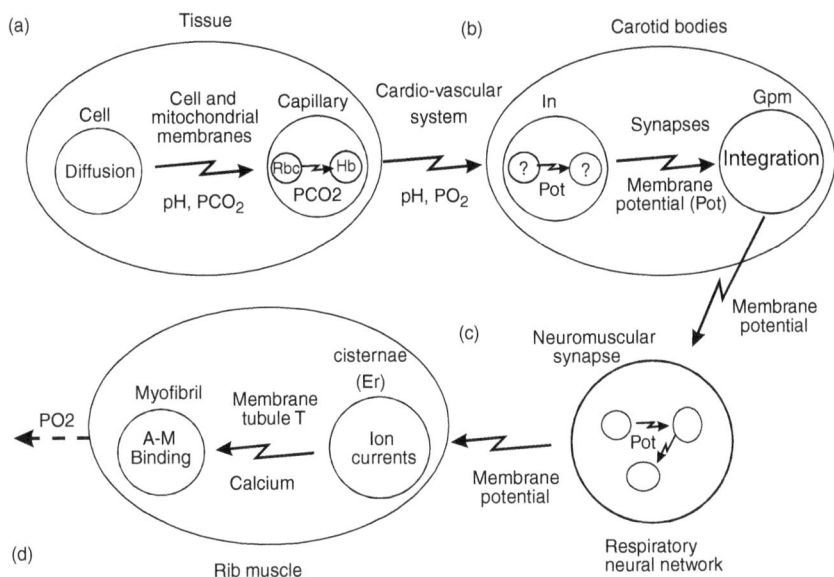

Figure III.4: An abstract representation of the functional schema of the regulation of "silent", cellular respiration. This regulation involves three types of control each linked to the activity of a specific receptor. (a) Acid by-products and CO_2 are transported to the tissue capillary; (b) The *neurochemical control* is effected by peripheral chemoreceptors situated in the carotid and aortic bodies; (c) The *chemical control* is provided by receptors located in the medullar region of the brain stem; (d) The *neuromechanical control* depends on mechanoreceptors situated in lung tissue and the thoracic wall. Rbc = red blood cell; Hb = hemoglobin; In = interneurons; Pot = potential; Gpn = glossopharyngeal nerve; Er = endoplasmic reticulum; A = actin; M = myosin.

transports 1–5 liters of oxygen per minute to the various tissues, and an equivalent quantity of carbon dioxide away from the tissues. The ventilation, a mechanical phenomenon involving muscular activity, must therefore be modulated according to the needs of the body. How is this done? When the ATP requirements of a body increase, the oxygen concentration in the cells decreases while the carbon dioxide concentration increases. The extracellular medium reflects these changes by the variations of the partial pressures of these gases in the blood. Similarly, the variations in the concentrations of these

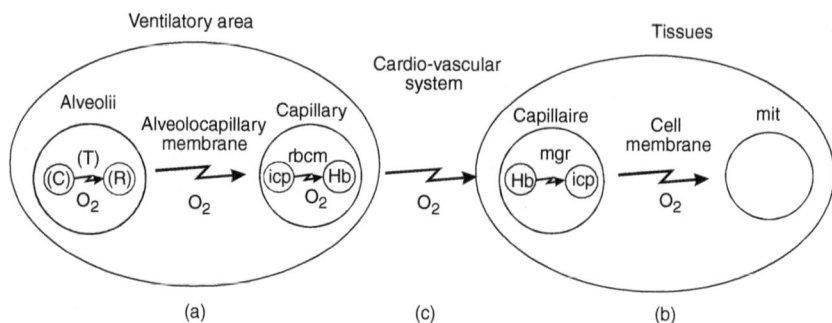

Figure III.5: Functional respiratory organization. Here, the functional interaction is the partial pressure of oxygen (PO_2), which represents the concentration of the oxygen transported: (a) from the ventilated alveolus in the conduction zone (C) to the respiratory zone (R) *via* the transition zone (T) (from the intracapillary plasma (icp) of the pulmonary capillary to the hemoglobin (Hb) *via* the red blood cell membrane (rbcm)) *via* the alveolocapillary membrane; (b) from the capillary in the opposite direction, and from the intracellular mitochondrion (mit) *via* the cell membrane; and (c) from the ventilated region to the target tissue *via* the cardiovascular system.

gases may be monitored in the cardiovascular system, in the tissues of the lung, in the lung capillaries, in the alveoli, in the red blood cells, and in the hemoglobin molecules — in short, in all the structures of the respiratory system. Figure III.5 gives a formal view of the hierarchical functional interactions of the respiratory system.

Let us consider the different steps of the respiratory function. The first step is the *conduction of air* collected from the environment and directed to the alveoli; this is followed by *alveolocapillary exchanges* during which oxygen moves from the alveoli into the capillaries while carbon dioxide moves in the opposite direction (Fig. III.5a). Once the oxygen crosses the alveolocapillary barrier, two more steps are required to ensure delivery of oxygen to the target tissues. First, the alveolar oxygen binds rapidly to hemoglobin, enabling the *massive transport of oxygen* (70 times as much as would be carried in the ordinary dissolved form). This is followed by *blood-tissue exchanges*, facilitated by the reversibility of the binding of oxygen to hemoglobin, on one hand, and the level of the blood pressure, on the other (Fig. III.5b). Simultaneously, the carbon dioxide collected from the tissues is released in the lungs by an analogous process of rapid dissociation of the gas from the tissues followed by binding to hemoglobin.

The regulation of the respiratory function is schematized in Figure III.4. The respiratory neurons (Fig. III.4c) emit an action potential that is transmitted to the respiratory motoneurons in the spinal cord. At the neuromuscular synapse, calcium ions are expelled from the cisternae of the muscle fiber and migrate towards the myofibrillae (Fig. III.4d), where the interaction between the actin and myosin molecules then produces a contraction of the muscle fiber. In this representation, the functional interaction is constituted by the membrane potential transmitted from the network of respiratory neurons to the muscle fibers of the thoracic muscle, in which the functional interaction is represented by the transport of calcium ions from the cisternae to the myofibrillae.

In the brief description given here, the regulatory mechanisms of the respiratory function have been deliberately simplified. The main object has been to present respiratory regulation as a subsystem of a major system related to the other systems of the body, not only the brain, or the limbic system that deals with the emotions, but also the renal and the cardiovascular systems. The complex relationships involved should incite us to try to understand the principles of the functional order underlying the general organization of the body.

The two major equilibria of the internal medium discussed above clearly reveal the difficulties raised by any approach to *integrative physiology*. The coupling between the components of the physiological functions may produce unexpected effects.[32]

---

[32]These effects, or *actions*, include the elaboration of "*products*" in biochemical pathways.

We may mention some striking examples. The central nervous system influences the hypothalamic osmoreceptors, so that a sudden emotional shock, for instance, might block the victim's breathing. The digestive apparatus controls the local extracellular volume through the vagal nerve under the control of the central nervous system, a local innervation, and two hormones: kinin and cholecystokinin; thus a state of anxiety could lead to the development of an ulcer. The cardiovascular system, acting through variations in the blood pressure and the cardiac flow, powerfully influences all the regions of the body, including the brain; intense physical exercise may therefore accelerate the cardiac rhythm and decrease cerebral blood flow, eventually producing a loss of consciousness. The microcirculation, under local as well as remote control, regulates transcapillary fluid exchanges, in particular the filtration of water and solutes in the kidneys, so as to maintain the hydro-electrolytic equilibrium of the body. The renal function also plays a fundamental role in maintaining the acid-base equilibrium through the transport of potassium and calcium ions. The relationship between these regulations, similar to that existing between the pulmonary and renal functions, is strongly dependent on the general metabolism of the cells and on the microcirculation.

> The main problem may now be clearly stated: how can we formulate a functional theory to describe such a complex set of physiological functions? Hitherto, we have merely been able to consider subsets of these functions. The full integration of the physiological functions of the living organism calls for a theory of functional organization. Moreover, such a theory would have to be a theory of action based on the hierarchical nature of the organism considered as a dynamic system, and on the functional relationships between the different constitutive elements.

Is the nature of the biological universe mathematically intelligible?

Could any reasoning based on the use of ordinary "verbal" language alone lead to a satisfactory explanation of the biological systems observed? This seems rather unlikely since such complex systems could hardly be understood without a formal theory, stated at least in terms of symbolic logic, to clarify our reasoning. In this sense, as in the case of the physical universe, the intelligibility of the biological universe would appear to be mathematical in essence. In fact, mere verbal reasoning proves insufficient in dealing with the intricacies of the normal physiological functions of living organisms, not to mention the physiopathological phenomena encountered in medical practice. To be sure, professional experience may enable physicians to identify, through analogy and deduction, the factors influencing the course of certain diseases. Thus, in Claude Bernard's time, most of the effort was directed, on one hand, to collecting and classifying medical facts so as to determine the causes and consequences of diseases and, on the other, to the use of the experimental method to obtain a satisfactory interpretation of the observations. Current techniques appear to offer much more — and at the same time, oddly enough — much less than the nineteenth century approach. Much more, since the accumulation of details over the years provides ever-deeper insights into the local spatiotemporal aspects of the phenomenology, and much less, since precisely because of this progress, local investigations tend to be increasingly compartmentalized. If all living organisms share a characteristic, it is surely that of being fundamentally indivisible and therefore irreducible to their separate, component parts. It is therefore encouraging that new methods based on the theory of graphs and the dynamics of systems are now finding their way into the teaching of physiology. (See, for example, the representation of physiological functions by the logical diagrams constructed by Silbernagl and Despopoulos, 1992.)

The living organism is essentially of a global nature and, although it can be fragmented just as inanimate matter can be physically broken down, we all know that, in general, it will not be able to survive. Although inanimate matter does possess certain global properties, such as the dislocations observed in crystals, a physical object may be studied even after considerable breakdown. The usual physical laws will hold good for the fragments, and the Coulomb forces or the van der Waal forces, for example, will continue to apply as such, or with only minor adjustments. The reason behind this lies in the fact that the local properties of inert matter are governed by the great principles of physics. This is clearly not the case for living organisms. In contrast to the local properties of inanimate matter, living organisms possess global properties, as revealed by the presence of structural discontinuities. It is true that molecular biology has recently proved successful in establishing the relationship between cause and effect in genetic diseases. Nevertheless, in spite of the powerful techniques now available, the correspondence between the structure of the genes and the functional properties of the organism is far from being established. And it may prove to be even more difficult to determine the origin of diseases involving the fundamental mechanisms of physiological integration that are so closely linked to the homeostasis of living organisms. Cardiovascular disorders, mental illness and cancer are among the diseases belonging to this category, and clearly only a global approach, integrating the underlying functional interactions from the molecular level to that of the whole organism, could be expected to yield new physiological insights.

IV

STRUCTURAL AND FUNCTIONAL ORGANIZATIONS OF LIVING ORGANISMS

A New Look at the Structure–Function Relationship in Biology

IV

STRUCTURAL AND FUNCTIONAL ORGANIZATIONS OF LIVING ORGANISMS

A New Look at the Structure–Function Relationship in Biology

Let us look back briefly on the preceding chapters. In Chapter I, we defined the functional organization of a living organism as a set of interactions represented by a mathematical graph constructed in accordance with the observed physical structures. In Chapter II, we presented a variety of examples of biological systems, and deduced the concepts of structural discontinuity, non-locality, structural hierarchy, functional hierarchy and specialization. We touched upon the existence of a causal relationship between structural discontinuity and the property of non-locality, probably due to the phenomenon of specialization in living organisms. In Chapter III, we showed that the homeostasis of the functions of a living organism is the result of highly regulated interactions between local transformations in biological structures, which are necessarily situated at a distance in physical space. We saw that the internal functional equilibrium of a biological system is maintained by the action of cybernetic feedback loops. We now propose to extend the idea of this elementary structure–function relationship to the global relationship between the structural and functional organizations of the living organism. This calls for some caution, since the concepts of structure and function do not have the same significance for the epistemiologist, the physicist or the biologist, because of the different approaches inherent to their respective disciplines.

Structure and function from the epistemiological, physical and biological points of view

From the epistemiologist's point of view, the concepts of structure and function have to be integrated within a unified scientific framework that is coherent and general. However, the meaning of the term "structure" has changed with time. Formerly, it was used rather restrictively to refer to the spatial disposition of elements, whereas today it is often used in a more general sense to indicate the relationships between the elements of a system, be it abstract or material. To avoid ambiguity, we propose to use the term "structure" for the *static* arrangement of elements, and the term "system" for the *dynamic* interactions of its elements. In particular, we shall adopt the definition of a system, equally applicable in mathematics and in biology, as given by Delattre (1971a):

> "A system is a set of interacting elements that may be distributed in a finite — or infinite — number of bounded classes by applying the laws of equivalence. The interactions between the elements can then be represented by the relationships between these classes in terms of the topology, the order, and the transfer of elements. Specific axioms concerning the conditions for the existence of the elements may then be enunciated for each system."

From the physicist's point of view, the structure of an object or, more generally, the structure of matter, corresponds to sets of particles, atoms and molecules bound together by interacting forces. Thus, the investigation of the structure of matter calls for the use of high-energy accelerators to break down the atoms. This is followed by the reconstitution of new sets of molecules by chemical techniques. Finally, these molecules are studied to determine the physical principles underlying the stability — or the instability — of matter. Thus, the significance of the physical notion of structure is reasonably clear.

From the biologist's point of view, the concept of structure is inextricably linked with that of function. For this reason, the general definitions of epistemology, no matter how precise they may be, will have to take into account not only the *biological nature* of the object studied but also the *biological method* used. Biological organization has three major characteristics. First, the definition of a given biological structure depends upon experimental constraints since it must be definable according to morphological, or at least quantitative, criteria. Thus, a ribosome observed inside a cell, the cell itself, and the tissue to which the cell belongs, all constitute distinct structures that can be observed and analyzed. The associated functions are often intuitively defined as the set of activities or transformations subtended by the corresponding structures. Secondly, the anatomical and histological hierarchy does not coincide with the hierarchy of the physiological functions or sub-functions. And thirdly, the higher we rise in the anatomical or functional hierarchy, the smaller is the number of elements involved, tending to the order of unity since the organism itself is at the highest level of biological organization.

To be coherent with the definition of the function, *the structure will be considered as the spatial, "physical" arrangement of the elements of the system.* In this sense, the structure will not be identified with the *relational* structure of the system but will coincide with the histology and the anatomy of the system such that the structural hierarchy corresponds to the hierarchy of the material system observed. We have already considered two essential characteristics of biological systems, i.e. the complexity of the functional interactions, illustrated by the typical examples of acid-base and hydroelectrolytic homeostasis (Chapter II); and the distinction between the functional and structural hierarchies of a living organism (Chapter III). The understanding of the inter-relationships between these hierarchical systems and the complex integration of the physiological functions of the body calls for a coherent representation of a living organism. Thus, our main objective will consist in constructing an "operational", mathematical definition of a generalized physiological function.

IV

The importance of stability in biological phenomena

The problem biologists would like to tackle is that of the integration of physiological functions in living organisms. But as we have seen, in spite of the powerful methods of calculation now available, the quantitative, analytical approach has failed to produce satisfactory results. An interesting possibility seems to lie in the use of a qualitative approach in the identification of the general principles underlying biological systems. These principles may be expected to reveal a functional order that could then be further investigated quantitatively. Thus, within the framework of an appropriate theory, i.e. with a suitable mathematical representation, we should be able to study the stability of biological systems. Let us recall that from a mathematical point of view, the stability of a system merely expresses the fact that the system tends to return to its equilibrium position after it has been "slightly" displaced. However, what precisely constitutes a "slight" displacement remains to be defined.

Dynamic systems possessing the property of stability belong to a special category, called *dissipative systems*. In these systems, a displacement from the equilibrium position is followed by a series of damped oscillations that diminish in amplitude until a resting point, called the *equilibrium point*, is reached. If the dissipative system is initially not in its equilibrium position, it will return to the equilibrium point with a damped oscillatory movement tending indefinitely to equilibrium.

The well-known example of the simple pendulum allows us to grasp the distinction between *stable* and *unstable* equilibrium. If the pendulum is released when it is in its vertical position, with its weight directly below its axis of rotation, it will not move. If released with the weight in some other position, it will oscillate and finally come to a stop in its equilibrium position. This corresponds to the *stable equilibrium* position. However, another position of the pendulum — with the weight directly above its axis of rotation — also corresponds to a position of equilibrium, but this is actually a position of *unstable equilibrium* since even the slightest displacement will cause it to perform a series of oscillations before it reaches its position of stable equilibrium. Systems of this kind are called *dissipative* because the energy of such systems is dissipated by friction into the surrounding medium. This irreversible change takes place at the expense of the kinetic energy of the system so that the movement of the pendulum stops after a certain time.

Another example, already mentioned in Chapter I, will here allow us to illustrate the classic notion of the *potential* of a system. In a mountainous region (Figure IV.1), rainwater will naturally run down mountain slopes and collect in lake areas in the valleys (Figure IV.1a). In cases of overflow, the water from a lake at a higher level will flow into a lake at a lower level (Figure IV.1b). Now, if there are two lakes on different sides of a mountain peak (Figure IV.1c), a slight difference in the spot where the rain falls at the summit will cause the water to flow into one lake or the other. In this example, there are four possible positions of equilibrium, two of which are stable, and two unstable.

Lines drawn through the peaks, ridges and valleys of the region represent the relief of the landscape. This relief is called

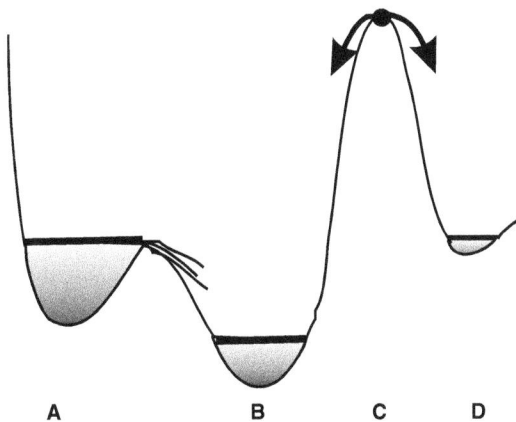

Figure IV.1: A vertical section showing the relief of a mountainous region in which water from lake A flows into lake B. The summit C represents a point of unstable equilibrium.

the *potential* of the system. The value of the potential at a given point corresponds to the altitude at that point, so that the highest point of the region has the maximum potential, and the lowest point of the lowest lake has the minimum potential. This description in ordinary language is equivalent to the mathematical statement according to which stable equilibria correspond to values of minimum potential. Thus, instead of saying that water flows down slopes, we may say that the potential decreases along the trajectories. In other words, unless it is in equilibrium, the state of the system changes with time such that its potential decreases at each successive instant. This defines the irreversibility of the dissipative system.

We may therefore represent a dissipative system by means of contours and equilibrium points as in Figure IV.2. The contours are curves formed by joining points with the same potential so that lines perpendicular to the contours constitute interesting trajectories inasmuch as they are the lines with the greatest slopes. It is along these lines that the equilibrium points will be more quickly reached. If, for the moment, we leave aside the

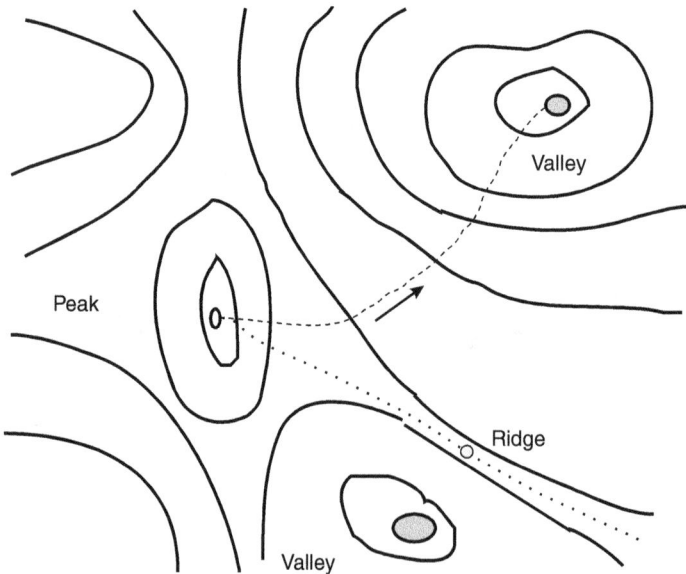

Figure IV.2: Contours of a relief map showing summits, valleys and ridges.

transitory phenomenon by which the dissipative system tends to its equilibrium point, then the dynamics of the system may be summed up simply by the initial and final states. The region in which all the possible initial states of a system converge to an equilibrium point is called the *attractor* of the system.

Figure IV.1c shows how the boundary of an attractor produces a delicate situation such that the slightest perturbation of an initial state on the boundary may swing the system from one attractor to another. Therefore, it may not always be necessary to determine the transitory phenomena occurring between the initial and final states of equilibrium. With complex systems, known to be dissipative, the knowledge of the equilibrium points and the attractors will often be found sufficient. This is the case in classic thermodynamics when the state variables of pressure, volume and temperature can be defined only at equilibrium. Without pushing the simplification so far as to ignore the transitory phenomena leading from a state of disequilibrium to one of equilibrium, we shall see that it is convenient to study a biological system in terms of the stability of its structural and functional organization.

The stability of dynamical systems may be studied by a method developed by Lyapunov (1857–1918). This mathematical technique reveals how such systems tend asymptotically to a stable state or even to an equilibrium point. However, there is no specific formula for actually determining a *Lyapunov function* for a given situation. Thus, only an intuitive approach will suggest an appropriate formulation that might be expected to lead to a good result. In the rest of this work, we shall be frequently using Lyapunov functions to develop our theory of integrative biology.

The physiological function viewed as a hierarchical system

We have defined biological "structure" as a physical, anatomical arrangement of material elements. In the preceding chapters, we defined various physiological functions, such as the respiratory and renal functions, on the basis of their specific action. However, from the mathematical or "operational" point of view, we still need a general definition that would be valid for all physiological functions. Thus, within the conceptual framework we have chosen, the physiological function may be defined by the set of interactions between the material elements, i.e. the structural units directly involved in the function. The function will then be represented by the solution of a dynamical system in which the equations describe the physiological mechanisms operating between the structural units of the system. This definition implicitly contains the *time scale of the process*, defined as the interval of time during which a significant variation of the solution occurs (see Chapter VI). For example, the neuronal activity in the nervous system varies over a time interval of the order of a millisecond, whereas the synaptic efficacy varies over much greater periods ranging from seconds to minutes. Thus, while we observe the membrane potential of a neuron, there may be practically no change in synaptic efficacy. Since the two functions, i.e. membrane potential and synaptic efficacy, operate on different time scales, we have a hierarchical classification of functions arising from the dynamics of the biological system. This is in fact the first observation of a temporal, functional order in a physiological system.

The dynamics of a system corresponds to the dynamics of the processes operating between the structural elements of the system. These processes are identified with the variation in time of the products exchanged between the structural elements localized according to their geometry, i.e. according to the histology and the anatomy of the system. For these reasons, we may say we now have an "operational" definition of the relationship between the structural and functional organizations of the system rather than a mere structure–function relationship:

A physiological function is defined as the global behavior of a physiological system, i.e. as the result[33] of the action of a set of

---

[33]The result of the physiological process may be an *action*, an *effect*, or a *product*, e.g. an electrical potential, synaptic efficacy, or the metabolic product of a biochemical pathway, and so on.

structural units belonging to different levels of the structural organization of the system.

With this definition of the physiological function in terms of functional interactions, we obtain an abstract construction — distinct from the real anatomical construction — represented by a set of parallel hierarchical functional systems. The hierarchy of the structural units, i.e. the anatomical hierarchy, which is well known in terms of molecules, organelles, cells, tissues, organs and organ systems, is linked to the existence of structural discontinuities. Our definition of the physiological function introduces a *functional hierarchy* based on structural units organized within hierarchical systems that do not generally coincide with the anatomical hierarchy. In practice, these systems are determined first by isolating a particular function, which is identified by the dynamics of the corresponding process and its time scale, and then by seeking the anatomical structures subtending this function. The same method will be applied to each subsystem obtained in order to determine the corresponding subfunctions, which will lead to the isolation of sub-subfunctions, and so on. The identification of the levels of functional organization depends upon the identification of the physiological subfunctions of which the spatiotemporal dynamics corresponds to the global behavior of the structural units at this level. Since this dynamics involves relationships between variables, the subfunction itself will be identified by these variables. We then deduce the following properties:

(1) *The abstract construction of the physiological function leads to a unique definition of the biological system.* At the lowest level of the structural organization, or the *fundamental level* at which auto-replicating molecules exist, the association of structural units leads to the production of complex molecules, i.e. proteins. These proteins constitute the essential elements of metabolic pathways that form interconnected networks. A given organ may produce several different types of molecule, for example specialized cells in the pancreas produce glucagon and insulin. In contrast, the activity of an operon will lead to the biosynthesis of only one

type of molecule. Thus, the specificity at the fundamental level will, by construction, lead to the unity of the physiological function as a hierarchical functional system. For example, the function corresponding to the regulation of glycemia involves the coupling of at least two physiological functions, i.e. the production of insulin and that of glucagon (Chapter II). These two functions are defined at the *same* organ level — that of the pancreas — but at *different* cellular levels, belonging to a lower level in the structural hierarchy.

(2) *Each structural unit is itself a hierarchical system organized at several levels, including the fundamental level.* Structural units that are similar according to certain histological and anatomical criteria constitute a *class of structural equivalence*. In mathematical terms, an equivalence class is a set of elements satisfying given relationships. For example, the *equality*, represented by the symbol: "$=$", satisfies the relationships of reflexivity ($a = a$), symmetry ($a = b$ implies that $b = a$) and transitivity ($a = b$ and $b = c$ implies that $a = c$). The equivalence class can then be considered as an element of a set on which algebraic operations may be performed. In the case of biological structures, the equivalence class corresponds to the geometrical identity of its elements. For example, the nephrons, which are identical anatomical and functional units of the kidneys, may be grouped into a single "nephron unit", the alveoli of the lungs may be grouped into a single "alveolar unit", and so on.

(3) *A level of functional organization corresponds to the activity of a set of structural units participating in the elaboration of a product[34] on a certain time scale.* It is represented by a mathematical graph (Chapter I) in which the summits correspond to the structural units and the oriented arcs correspond to the functional interactions. The product is

---

[34]The *product*, which results from the dynamics of the structural units, may be assimilated to the corresponding *physiological function*. More precisely, in the mathematical sense, the product is the value of the physiological function.

elaborated according to a given dynamics varying on a certain time scale. This defines the level of the functional organization.

(4) *An elementary physiological function is defined as the functional interaction between two structural units, i.e. the source and the sink.* The product resulting from this interaction may, in a certain sense, be identified with the elementary function. Thus, the physiological function can be viewed as the final result of a set of functional interactions that are hierarchically and functionally organized. Finally, the physiological function may be identified with the global behavior of the hierarchical system.

The hierarchy of functional organization

If we take into account the physiological mechanisms of the structural units, we see that each structural unit is in itself a hierarchical system. For example, the ribosome is a structural unit that may be compared to a chemical factory producing proteins, the mitochondrion is a structural unit that may be compared to a power plant, and so on. In the cell, the function corresponding to the production of enzymes includes the activity of two units, i.e. the unit corresponding to an operon, or a group of genes, and the unit corresponding to the ribosome, which eventually combines with other ribosomes to form the polysome (Figure II.3). These units constitute two parallel hierarchical systems. However, the two systems, i.e. that of the operon and that of the ribosome, are evidently not at the same hierarchical level. The ribosome is a complex structure at a very high level of organization compared to that of the operon.

To further illustrate the *relativity* of hierarchical levels, let us consider some other examples. One illustration is the direct interaction of hemoglobin with a neuron in the brain stem (Figure III.3). In this case, a structure at the cellular level of the central nervous system, i.e. the neuron, is associated with a structure at the molecular level of the respiratory system i.e. the hemoglobin molecule. Thus, the neurons of the brain stem and the molecules of hemoglobin will be situated at the same hierarchical level of the functional organization, although these structural elements belong to different levels of the structural organization. The functional hierarchy of the homeostasis of water and electrolytes (Chapter III) provides another interesting illustration. Thus, the tubules of the nephrons, which are structural units of the kidney, are at a certain structural level of the organ. These tubules, certain portions of which secrete hydrogen ions, interact functionally with carbon dioxide chemoreceptors controlled by the nervous system *via* the renal bicarbonate buffer system. In short, homeostasis can be understood only by considering the combined activity of two functional systems, i.e. the acid-base system and the hydroelectrolytic system.

These examples clearly show that the histological and anatomical hierarchy does not coincide with the hierarchy of the physiological functions or subfunctions. That this is generally the case may well have led to the confused arguments often put forward concerning the structure–function relationship. The difficulty appears to lie in the identification of the functional organization that implicitly depends on the dynamics of physiological processes and, consequently, on the formalized description of the phenomena involved. In the following chapter, we shall show how the functional hierarchy, based on the time scale governing these phenomena, actually emerges from a process of association between structural units taking into account the dynamics of the phenomena.

The structure of Pei's Pyramid at the Louvre Museum in Paris provides an interesting representation of a hierarchical system. Initially, the hierarchical system describing a physiological function may be envisaged as a set of nested systems, each system being contained within another, as shown in Figure IV.3. Each point, at any level except the fundamental level, represents a hierarchical system. Each element of a given system, considered at a certain level, itself corresponds to a group of elements at a lower level. However, this description of the hierarchy is insufficient since it fails to take into account the relationship between the structural and functional organizations. For this reason, we shall now introduce an original

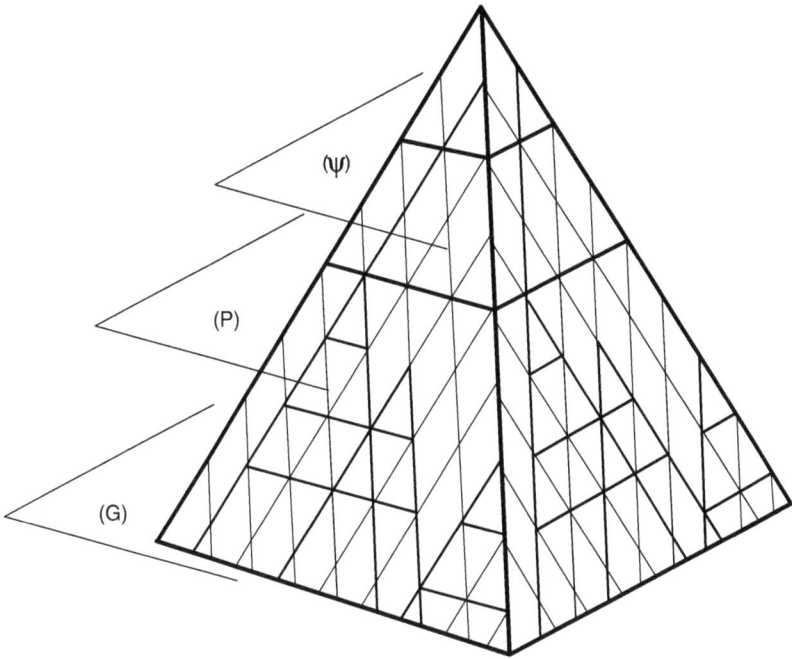

Figure IV.3: The structure of Pei's Pyramid at the Louvre Museum in Paris offers a good metaphor for a complex biological system. Each plane represents a certain level of organization. Each intersection between planes parallel to the faces of the pyramid represents a structural unit. The three planes shown on the figure, marked (G), (P) and (ψ), correspond to three levels of a physiological system involved, for example, in the production of a hormone (ψ). The epigenetic level and that of the metabolic pathway are represented respectively by (G) and (P) as described in Figure II.3 (see also Figure IV.5). The third level, (ψ), would correspond to the structural unit producing the hormone, for example, the thyroid gland in which thyroxin is elaborated by means of an interaction with another physiological system that synthesizes cortisol, i.e. another structural unit situated at the summit of another pyramid, either included in, or including, this pyramid (see also Figure V.4).

three-dimensional representation incorporating structural discontinuities, and leading to a new formalism.

From the mathematical point of view, the temporal hierarchy should allow us to consider the dynamics of a biological system as a set of dynamical systems decoupled with respect to time and dependent on variables and parameters, such that the parameters at a given level act as variables at a higher level.[35] However, from the practical point of view, we shall have to temper our enthusiasm since the complete description of the functional organization of a system assumes full knowledge of *all* the dynamics of the biological system and, consequently, *all* the elementary mechanisms underlying the phenomena observed. In the absence of such complete knowledge, we can only try to approach the final objective by means of a series of approximations that will have to be tested experimentally at each stage of the theoretical construction.

Three-dimensional representation of the relationship between the functional and structural organizations of a biological system

We may recall that a *structural discontinuity* in a biological system corresponds to the physical space separating two regions, each with specific functional properties associated with distinct physiological mechanisms. The simplest example of a structural discontinuity is that of a cell membrane separating regions in which different chemical reactions occur, e.g. the nuclear membrane (Figure II.4), and the inner or outer mitochondrial membranes (Figure III.4a). At a much higher level, the structural discontinuity may be represented by an entire physiological system, such as the cardiovascular system that transports a hormone, e.g. insulin (Figure II.1), or molecules of oxygen (Figure III.5), from a source to sink.

At a given level of the structural hierarchy, the structural discontinuity constitutes a structural unit at a lower level, for example the membrane with respect to the cell, the synaptic junction with respect to the neuron, the vascular tree with respect to the respiratory system, and so on. The physiological

---

[35]Contrary to the case of fractals, i.e. geometrical shapes that are generally *self-similar* at all levels of structure, the properties of the systems described here are not conserved from one level to another.

mechanisms operating at a lower level differ from those at a higher level.

Many examples of distinct structural and functional organizations are found even in complex animal and human societies. For instance, in a beehive, the hive made up of hexagonal cells represents the structural organization, whereas the specialized activity of the queen bees, the workers and the drones corresponds to the functional organization. The system as a whole, which exists essentially for the reproduction of bees, constitutes a functional hierarchical organization based on a structural hierarchy in which specialization plays a major role.

Again, the military architecture of the Middle Ages (Plates IV.1 and IV.2) clearly shows the relationship between the structural and functional organizations of the system composed of the fortified castle and its inhabitants. The castle, built of stone according to precise architectural rules, represents the structural organization, whereas the coordinated activity of the defending army corresponds to the functional organization of defense.

To be sure, the analysis of the complex structural and functional organizations of large societies does not appear to be feasible at present but we feel that an appropriate physiological representation of an individual biological organism is now within reach.

The physiological functions of a biological organism may be conveniently represented in a novel space called the *space of units* (Chauvet, 1993a), as illustrated by the view through our imaginary "selective" microscope (Figure II.6). More rigorously, a physiological function may be represented on the z-axis of a graph in terms of the hierarchical system of unit structural spaces distributed in physical space along the x-axis (Figure IV.4). Each level then corresponds to a structural discontinuity and is represented in the xz-plane by a straight line parallel to the x-axis that intersects the z-axis at a point. The shaded plane in Figure IV.4 illustrates an example of the structural organization for a certain physiological function (noted ψ_1) constructed on three levels of structure (noted 1, 2 and 3) corresponding to the spaces of units. According to the definition

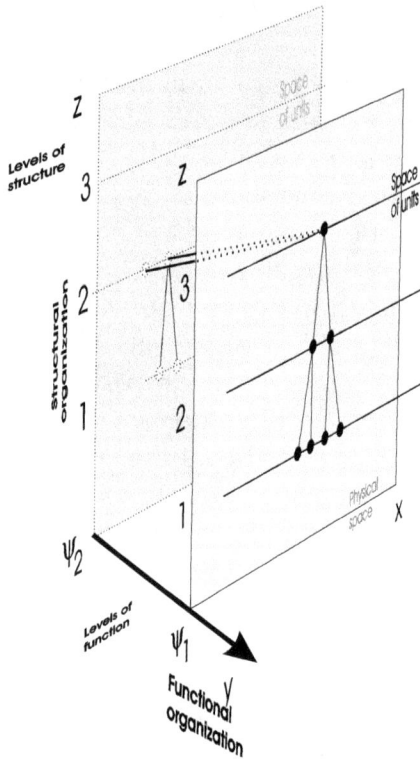

Figure IV.4: The z-axis and the y-axis correspond respectively to the structural and functional organizations of a biological system, whereas the x-axis represents the space of structural units. The lowest level of structural organization is the ordinary physical space in which the system exists. Each plane intersecting the y-axis represents a particular physiological function. For example, ψ_1 is a physiological function involving the structural units in this plane, classified according to their hierarchical structural levels (1, 2 and 3 in this illustration).

of the physiological function, all the physiological mechanisms operate on the same time scale at each of the levels. For example, in the case of the neuron, the synaptic channels open and close on the same time scale as the variations of the membrane potential. This means that a single dynamical system could be used to describe these processes by integrating the physiological mechanisms at the lower level of the synapses and the higher level of the neuronal cell bodies.

Let us now consider a somewhat more complicated case, for example, the activity of a group of neurons. This involves at least two elementary, interdependent physiological functions, i.e. on one hand, the spatiotemporal variation of synaptic efficacy and, on the other, that of neuronal activity, which itself depends on the synaptic efficacy. According to definitions given above, the synaptic efficacy and the neuronal activity are functional interactions between structural units. Since these physiological functions operate on different time scales, each may be assigned a specific level of organization. We thus have a double hierarchy, structural and functional, that may be represented by constructing a y-axis perpendicular to the xz-plane (Figure IV.4). Each graduation on the y-axis corresponds to a certain level of functional organization such that each level of functional organization has its own levels of structural organization. Thus, a physiological function may be conveniently represented by a three-dimensional graph. This is illustrated by the two examples given below, one taken from the nervous system and the other from a metabolic system.

In the nervous system, the variation of the membrane potential in neurons is due to two phenomena. One is the opening of "fast" channels on a short time scale, described by the synaptic efficacy ψ_1; and the other is the opening of "slow" channels (structures specific to the synapse that we have called the *synapsons* (Chauvet, 1993b) on a long time scale (Figure IV.4). This long time scale defines a second level of functional organization for the synaptic efficacy corresponding to ψ_2. Thus, the synaptic efficacy is defined for the set of structural units, i.e. the synapsons. The individual behavior of each synapson, i.e. the variation of conduction, results from the global behavior at the lower level of transmitters, receptors, channels, etc., on a certain time scale. The global behavior of the set of synapses, i.e. the variation of synaptic activity, defines the level of functional organization on another time scale. The set of synapses corresponds to the neuron as a structural unit. The set of neurons, in turn, defines the neuronal network as a higher structural unit, with a global behavior that, on yet another time scale, corresponds to its specific activity, which differs from that of the neuron. The same reasoning may be extended to structures ranging from the level of molecules to that of tissues. The description based on functional interactions allows us to take advantage of an implicit property. It is generally admitted that one neuronal network, i.e. a *nucleus* in the usual terminology of neurophysiology, acts on another in the nervous tissue, and one neuron acts on another in the neuronal network. However, it is less obvious that *one synapse may act on another in a neuron, or that one channel may act on another in a synapse.* The conceptual framework proposed here makes it possible to analyze these apparently very different types of action by means of a single formalism (Chapter VI).

In a metabolic system, a metabolite, i.e. the product of a biochemical pathway, is synthesized by the sequential action of several enzymes on substrates (Figure II.3). Each step of the pathway requires the action of an enzyme produced at the lowest level of organization (level 0). The set of substrates, which contributes to the product, defines another level of organization (level 1) so that each structural unit at level 1 contains level 0, or the fundamental level. Thus, we have a two-level *hierarchical system* as shown in Figure IV.5. To take this reasoning further, a third level of functional organization may be defined by the combination of the products of a set of biochemical pathways that would result in the "emission" of some final product. In other words, all the associated structural units that lead to the production of metabolites may be grouped together in a single structural unit. This structural unit, in which the functional interactions correspond to entire metabolic pathways, allows the elaboration of a complex new molecule. To illustrate this, the structural unit might be a *cell,* and the new molecule could be a *hormone* produced by this cell to act at a distance on another cell.

The dynamics at each level determines a specific time scale. Thus, cell proteins are synthesized on a much longer time scale than that of an enzyme reaction. Similarly, the variation of neuronal synaptic efficacy occurs over periods of the order of seconds or minutes, whereas the membrane potential may change in just a few milliseconds. Each global process, resulting from a set of partial processes, operates on a specific time

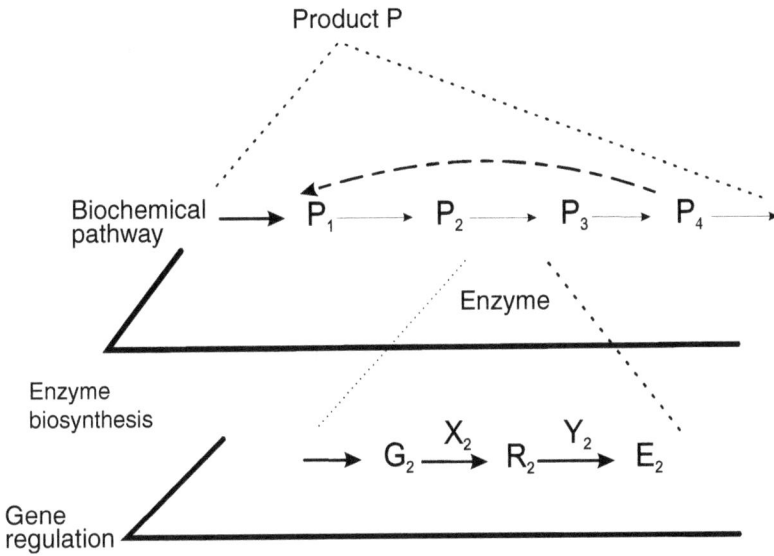

Figure IV.5: The hierarchical system of a metabolic pathway operating at two levels: metabolic and epigenetic. This illustration, which corresponds to the kinetic schema in Figure II.3, implicitly introduces different dynamics at the two levels of functional organization. The lower level is represented by the biosynthesis of an enzyme, noted E_2, by transcription of a structural gene G_2 followed by translation on a ribosome R_2. The enzyme E_2 catalyzes the transformation of P_2 into P_3. The set of steps involved in this process constitutes a metabolic pathway leading to the synthesis of a product P at a higher level of organization. In turn, the product P will be incorporated into another functional network at an even higher level of organization.

scale. Studies in chronobiology, i.e. the analysis of biological rhythms, suggest that the time scale of a given biological process is related to the level of organization of the corresponding biological system. Thus, the different time scales observed may be used to define a hierarchy of functions.

"Dual" representations of a biological system: (N,a) and (ψ,ρ)

Compartmental analysis offers a formalism often used in biology. It involves the abstract representation of dynamic phenomena in terms of the *occupation numbers* of a class, i.e. the number of elements in a class; and the *rate of transfer* of these elements from one class to another, i.e. the number of elementary transformations per unit time from one class to another. Compartmental analysis, which is associated with the notion of structural organization, differs considerably from our representation based on the combination of elementary functional interactions, which allows the theoretical construction of a hierarchical biological system.

Biologists have long favored the use of compartmental analysis for the formalization of observed biological phenomena, particularly since the introduction of radioactive tracers and other molecular markers that allow the determination of the fate of a given molecule in the body. For example, calcium exists in different physiological states, i.e. ionized or non-ionized, free or bound to other molecules in the blood, localized in bone tissue or in cell membranes, and so on. All the molecules in a given physiological state are considered to belong to the same compartment. In other words, *a compartment is defined as a quantity of homogeneous matter in which each element is in a given state at any given instant.* If several compounds are present in a biological medium, then each compound that is homogeneously distributed will define a specific compartment. The criterion of homogeneity is obviously restrictive since it excludes the investigation of transient states, but it has been found satisfactory enough in ordinary physiological work.

Thus, the compartment as defined above is not necessarily a volume in the physical sense, such as that of a cell, or a space in the physiological sense, such as the extracellular space. In most cases, there are no precise physical limits. The compartment differs from physical space although, of course, it exists *in* physical space. For example, in the blood circulation of a body, the matter made up of all the red blood cells in the same physiological state at a given instant constitutes a compartment. Similarly, calcium distributed in the bones, the blood and the cell membranes of a body constitutes another compartment. Compartmental models are used to describe the transformations of the *state* of matter in homogeneous, macroscopic subsets that constitute distinct compartments. Thus, the blood, the extracellular medium or an organ may be considered as typical compartments of the body. Delattre (1971a), who expressed the concept of biological compartments in terms of *classes of functional equivalence*, constructed a set of axioms generalizing compartmental analysis by means of a powerful *transformational formalism*. He suggested that the notion of a compartment, which is generally associated with space in the physical sense, should be replaced by a more relevant expression, viz. a *class of states*. Furthermore, according to Delattre:

> "The distribution of elements should be such that we always have a statistical system, i.e. a system in which all the elements belonging to a given class of states have exactly the same functions with respect to the other elements of the system. [...] The probability of a reaction occurring in unit time will then be proportional to the product of the cardinals of the species, each cardinal being raised to a power equal to the number of elements of the species participating in the reaction."

The statistical nature of the system requires, on one hand, that the probability of reaction be uniform for all the elements of a given class, and on the other, that there be a great number of elements (usually molecules) in each class. In effect, if the number of elements in a given class is sufficiently large, then the local temporal balance of the appearance and disappearance of elements will be given by a differential equation involving the number of elements in that class, and the number of elementary transformations per unit time from one class to another. Figure IV.6 illustrates the (N, a) representation, N being the occupation number of the classes, and a the "rate constant" of the transformation. Delattre's formalism of compartments and the more general formalism of transformation systems are applicable if the partition of the system into elements according to their functional definition allows the construction of *classes of functional equivalence*, and the number of elements in each class is large enough to produce a statistical system.

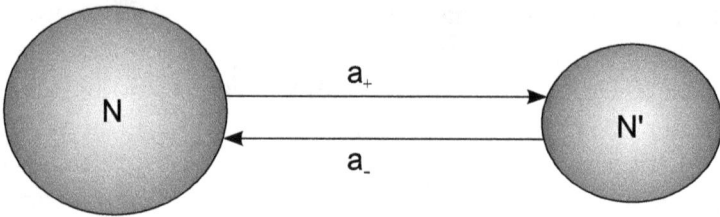

Figure IV.6: Transfer between two compartments having rate constants a_- and a_+.

In spite of its rather abstract nature, compartmental analysis has been widely used in biology since many of the systems of the body can be mathematically described by linear differential equations.[36] The solution of these equations is given by the sum of the exponential functions that are determined by the analysis of experimental curves. This allows the determination of the numerical values of the parameters of the exponential functions, i.e. the rate constants of the exchange between compartments. The quantification obtained is specific to the system investigated and allows an interpretation of the phenomenon observed. In fact, the use of compartmental models gives good results when applied to biological, and even ecological, processes, if the systems can be partitioned according to the interacting substances in the compartments. Since each compartment is assumed to be homogeneous, the rate at which the elements are mixed in the compartment must be far more rapid than the rate at which the elements are transferred from one compartment to another. *Thus, a mathematical spatiotemporal relationship is implicit in the (N,a) formulation.*

However, if functional interactions are taken into account, the (N,a) representation can no longer be used. As we saw in Chapter I, a functional interaction, with its properties of non-symmetry and non-locality, corresponds to the transport of a product, noted ψ, from one point in space to another at a finite rate. The functional organization is thus based on structural

---

[36]The simple compartmental structure shown in Figure IV.6 leads to the system of differential equations: $dN/dt = a_-N' - a_+N$

units that are, as we have seen, classes of structural equivalence. All the units involved in the emission, i.e. the dynamics, of the product may be gathered into a single structural unit at a higher level of organization. Then, if the spatial density of these units, which is generally not uniform, is noted ρ, *the* (ψ, ρ) *representation may be used to describe the functional organization.* Since the product is transported from one unit to another, i.e. from one point in space to another, this representation incorporates the geometry of the system.

As opposed to the (ψ, ρ) representation, the (N, a) representation assumes that the product is homogeneously distributed in space, constituting a compartment. In other words, the (N, a) representation is essentially associated with the structural organization of the biological system. This evidently constitutes an important difference between the two representations. Another fundamental difference appears when we consider increasingly higher levels of organization since the number of elements in each class progressively decreases towards a very small number indeed at the level of the organs. Clearly, a statistical system, which requires that each class has a large number of elements, will no longer be applicable.

Physiological studies generally make no distinction between *substrates* and *actions* since these categories do not fit neatly into any one representation, i.e. (N, a) for the substrates, and (ψ, ρ) for the actions. This may be illustrated by the function of oxygenation of tissues by the respiratory system (Figure IV.7). In the (ψ, ρ) representation, oxygen molecules are transported from one structural unit, i.e. the bronchi, to another, i.e. the tissues. In the (N, a) representation, the changes of state occur at a certain rate between the compartments composed of alveolar oxygen, capillary oxygen and tissue oxygen. The rates of transfer between the compartments are, in general, constant quantities that give a simple description of the changes of state.

Let us consider a compartment in the (N, a) representation. If a represents the velocity at which the elements leave this compartment, then the variation of N, the number of elements in the compartment, will be a times N per unit time. From the

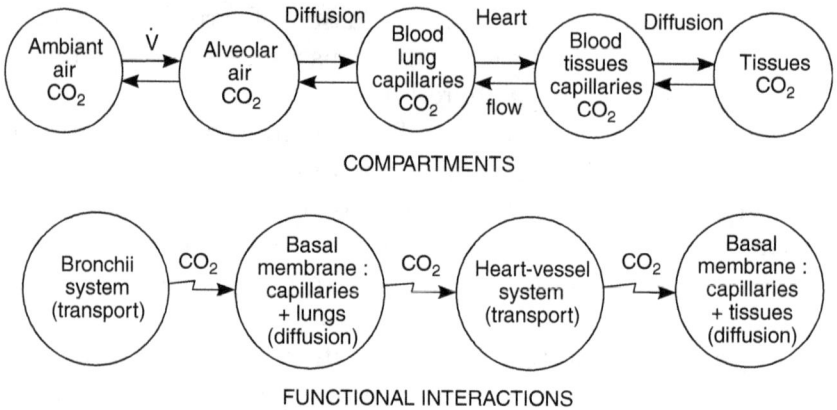

Figure IV.7: Correspondence between the representations of the respiratory function in terms of compartments, on one hand, and actions (or fields), on the other. The transport of oxygen molecules corresponds to a functional interaction in the (ψ,ρ) representation, whereas the ensemble of oxygen molecules constitutes a compartment in the (N,a) representation.

phenomenological point of view, this velocity implicitly takes into account a large number of factors, such as the geometry of the system and the operation of local mechanisms. It gives a macroscopic description of the "average" behavior of the global system composed of a set of compartments. The simplicity of the expression is due to the statistical nature of the system in which a very large number of molecules is involved. Thus, the variable a implicitly includes, on one hand, the geometrical properties of the anatomical structures subtending the compartment, represented by the density ρ, and on the other, the local mechanisms represented by the variable ψ. *The ψ and N representations may be considered "dual" representations.* Here, the term "dual"[37] implies that, in the (ψ,ρ) representation, ψ describes the matter transported between structural units with a distribution of density ρ, whereas in the

[37]In mathematical terms, two expressions are said to be *dual* if each can be deduced from the other. Thus, by analogy, the duality of the representations (ψ,ρ) and (N,a) means that: $\psi = f_1(N,a)$, $\rho = f_2(N,a)$, $N = g_1(\psi,\rho)$, $a = g_2(\psi,\rho)$.

(N,a) representation, N is the number of elements in the compartment in a given state. The factor a is a rate constant since it describes the way in which the transfer is carried out between two compartments. The variables that correspond to each other in the dual relationship are ρ and N, on one hand, and ψ and a, on the other. It is rather difficult to express this transformation in explicit terms since it involves the ensemble of states of the elements in the compartment and these are in fact implicitly included in the process ψ, which is a function of time. The transformation is thus a non-explicit relationship between three of the variables. This is why the choice of the representation depends on the problem posed. When the number of units in the compartments is large, for instance in the case of a molecular system for which we wish to determine the variation with time from one state to another, compartmental analysis would be perfectly appropriate.

Some consequences of the (ψ,ρ) representation

The (ψ,ρ) representation, which indicates the *totality* of the dynamics of a system in terms of functional interactions, could be expected to give a clear image of the functioning of the system. However, the complexity of the representation increases rapidly with the number of levels of organization and the number of systems in the organism. The multi-level graph becomes so inextricably complex that it can only be analyzed by using powerful computational methods. The source-to-sink relationships can be organized by means of a special computer program such that the hierarchy of the system is automatically displayed on a screen. Then, by selecting out a particular structural unit we may explore the different hierarchical levels and move from one unit to another *to see exactly where a perturbation occuring in a certain part of the body produces its final effect.* Clearly, this type of interactive graph opens up promising new lines of investigation in integrative physiopathology.

The phenomenon of homeostasis (Chapter III) offers a good illustration of the notion of integrative physiopathology. The regulation of the hydroelectrolytic and acid-base equilibria

IV

involves all the major systems of the body, i.e. respiratory, circulatory, digestive, excretory, and neurohormonal. A perturbation in any of these systems could affect the cardiovascular system and produce a state of *heart shock* characterized by widespread cell suffering, accompanied by signs of total disequilibrium in the body, leading rapidly to death. Clinical observations of this condition reveal a *vicious circle* comprising the following sequence of events: constriction of blood vessels, decreased oxygenation of cells, local acidosis, local edema, decreased blood volume, decreased venous return, heart failure, and generalized acidosis. In fact, the state of shock is aggravated by the metabolic, cellular and tissue disorders that increasingly damage the lungs, the digestive tract, the kidneys, the heart and the brain. This brief description of heart shock suggests how difficult it might be for a physician to interpret the clinical and biological results in such cases. In practice, all the physiological functions of the organism are so seriously perturbed that it may be extremely difficult to analyze the syndromes. Since the vicious circle leads spontaneously to death, it would be important to determine the critical point, i.e. the crucial instant beyond which all the essential organs will be irremediably damaged.

Now let us see how an interactive graph could lead to new insights into the mechanism of the state of heart shock. Figure IV.8 shows a *cyclic sub-graph* that emerges from the interactive graph of the functional organization of the body. The cyclic sub-graph corresponds to one of the pathways of the main graph in which, by moving from one summit to another, we return to the starting point. A more detailed graph, taking into account a greater number of levels of organization, would show a greater number of interactions in the form of cyclic and non-cyclic sub-graphs. This would of course provide a fuller explanation of the mechanisms underlying the origin of the state of shock and its consequences.

Thus, an abstract, mathematical representation of biological reality could lead to a better understanding of the problems posed by integrative physiology. *In fact, the topology of the organism indicates the <u>localization</u> of the alterations caused by*

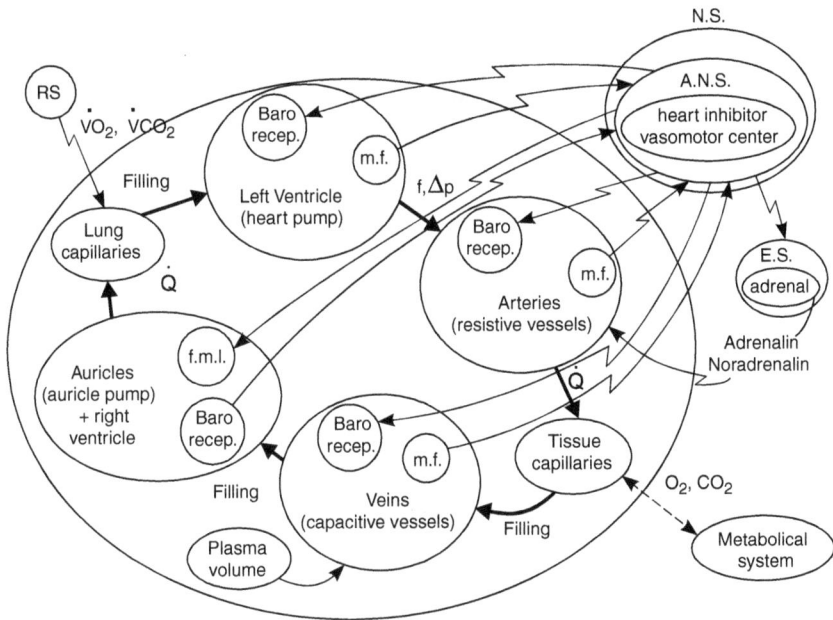

Figure IV.8: The cardiovascular system is represented here in terms of actions, or fields. The vicious circle of the state of shock appears in the form of a cyclic sub-graph of the functional interactions in the (ψ, ρ) representation. RS = respiratory system; NS = nervous system; ANS = autonomous nervous system; ES = endocrine system; m.f. = smooth muscle fiber; f = respiratory frequency; Δp = pressure gradient; $\dot{Q}$ = blood flow; $\dot{V}_{O2}$ and $\dot{V}_{CO2}$ = ventilatory flow of oxygen and carbon dioxide respectively.

the perturbations; and the dynamics of the processes involved, which is related to the geometry of the organism, describes the nature and the course of the resulting alterations. As we shall see, the time-variation of a biological system is expressed by two dynamics: one corresponding to the time-variation of its organization (Chapter V), and the other to the time-variation of the processes taking place within this organization (Chapter VI). We propose to deduce a classification of functional interactions and structural units according to the specialized hierarchical physiological systems to which they belong. The graphical visualization of the structural and functional organization of the organism would then allow us to identify particular pathways, such as the cyclic sub-graph in the case of the state of shock discussed above. The determination of the path followed by a perturbation from a given point of the graph under certain constraints should predict the consequences of the perturbation in the organism. The construction of a "functional graph" should lead to an understanding of the role played by functional interactions and establish the "best" topology of the system according to selected criteria, in other words it should allow us to verify whether the constraints actually observed correspond to an optimum organization of the system. Finally, we shall see how the *potential of functional organization* for certain physiological functions may be determined.

It might be held that such an ambitious project is bound to be illusory. The very idea of deducing biological diversity from general principles may appear unrealistic or even illogical. However, the biological diversity we propose to investigate is not the diversity of the form or the structure of living organisms. What we are interested in is the origin of the functional characteristics that are common to all forms of life. Thus, our main objective is to construct a representation that expresses these characteristics adequately. Obviously, this is bound to be a long and difficult task, full of dangerous pitfalls. But efforts in this direction may be expected to lead to a better understanding of the topology of a living organism and give new insights into the physiological functions observed, including the effects that perturbations might produce. We may even hope to explore the possibilities of evolution of various biological species by comparing their potentials of functional organization.

The theoretical representation of a living organism

The problem posed by the theoretical representation of a living organism is somewhat reminiscent of the ancient philosophical quest for the explanation of life. The description of physical reality is based on the use of a few general principles to create mental constructs that can then be suitably assembled to establish a physical theory. However, biological science has not yet acquired the general principles that would allow us to conceive an abstract image of the functioning of a living organism. To be sure, we know how to model the workings of certain organs of the body, such as the lungs, the kidneys or the heart, but we still have no way of integrating these models into a general biological theory. A possible theoretical approach to the functional aspect might lie in the representation of living organisms in terms of combinations of functional interactions between hierarchical systems. If each of the interactions satisfies dynamics that can be deduced from established biological laws, then the set of interactions might be integrated into a system governed by just a few extremum principles.

An extremum, or optimum, imposes a mathematical condition on a certain state function of the system. This global condition ensures that, at each instant, the system will be maintained on an optimum course. Such a condition may be natural, internal to the system, or imposed externally. For example, in physics, the principle of least action (Chapter I) is an optimum principle since, in its simplest form, it states that the movement of a conservative system in space is such that the average difference, in the course of time, between the kinetic and potential energies of the system remains at a minimum. A mathematical method, called *variational calculus*, may then be used to calculate the optimum trajectory of the system.

As we shall see, the theory of biological systems presented in the following chapters, leads to an optimum principle. In this theory, *the physical structure of a living organism, composed of molecules stabilized by symmetrical, interacting forces, will be associated with a set of hierarchical functional systems stabilized by non-symmetrical and non-local functional interactions*. Since Newton first introduced the abstract concept of force in physics, we have learned that the action of a set of forces in a physical system, i.e. in its structural organization, leads to a global behavior. This behavior is governed by an optimum principle if each of the forces satisfies, at the elementary level, the dynamics of certain laws, such as that of molecular attraction. The ineluctable, irreversible tendency of a dissipative system to move towards its equilibrium point is an example of an optimum principle since each stable state of the system is at a minimum potential. This is observed in physics, for example in a thermodynamic system that tends to conditions of uniform temperature. However, does the same type of reasoning apply in the case of a living organism when we consider the dynamics of elementary functional interactions in a hierarchical system that differs from the structural system? In other words, is the functional organization of a biological system, which we have shown to be distinct from its structural organization, also governed by optimum principles?

It is rather difficult to believe that the visible functional organization of the structural units of a living organism is the result of mere chance. This extraordinary organization must surely be subtended by some general principles, just waiting to be discovered. In the next chapter, we shall present two original hypotheses that might, after adequate experimental testing, be acceptable as the basic principles underlying the unity of a biological system as a *hierarchical system*. The first is the *hypothesis of self-association*, which determines the *existence* of the functional interactions of a biological system, and ensures the topological stability of their organization. The interactions are such that the corresponding domain of stability is actually extended. This phenomenon, almost never observed in physical systems, suggests that *biological stability is related to complexity*. The second is the *optimum* hypothesis, which defines a biological system as one having the largest potential of organization. This hypothesis is at the origin of the *hierarchical organization* of functional interactions, the role of the organization being to increase the functional order of the biological system.

If these hypotheses are found acceptable, *a biological system may be considered self-organized such that it extends its domain of stability and simultaneously increases its functional order*. This important result, if experimentally verified, would prove that the levels of organization observed in living organisms satisfy a vital necessity for their existence, i.e. the maintenance of their physiological functions. Thus, among all the functional organizations theoretically possible, the observed functional organization would be the *unique* organization satisfying the two fundamental properties,[38] i.e. the extension of the domain of stability, and the increase in functional order, which in conjunction ensure the invariance of the physiological functions of a living organism. As we shall now show, these properties allow us to deduce the *invariants* of the functional organization necessary for the permanent functioning of the biological system, analogous to the invariants of the structural organization that ensure the permanence of the structure of a living organism.

---

[38]Mathematically, as we shall see in the next chapter, these properties are of a *variational* nature.

V

PHYSIOLOGICAL CONSTRAINTS OF BIOLOGICAL DEVELOPMENT

Towards An Evolutionary Principle of Life

V

PHYSIOLOGICAL CONSTRAINTS OF BIOLOGICAL DEVELOPMENT

Towards An Evolutionary Principle of Life

Could the dawn of human consciousness have been associated with the perception of the beauty of nature? The sensations and emotions produced by the shapes, colors, textures, flavors, sounds and movements of all the elements of the universe — the great and the small, the living and the non-living — have surely inspired artists over the ages. Scientists, too, have found beauty in the investigation of the inanimate and the animate world. Thus, physicists have discovered exquisite relationships between matter and energy, while biologists have been enthralled by the glorious mysteries of life. However, is the living organism not an integral part of the non-living universe? Do we differ in some essential way from the world we live in?

Although these major philosophical problems are far beyond the scope of this work, we might try to address some aspects of evolutionary biology. For instance, how do we explain the existence of the functional biological interactions observed during the evolution and development of living organisms? In other words, what is the origin of the privileged relationship between certain structural units of a living organism during its evolution and development? It is surely extraordinary that the complexity of organisms should actually *increase* not only in the course of evolution from one species to another, for example, from the earthworm to the mammal, but also during the development of an individual organism from the molecular level to that of the tissues and organs. Clearly, a biological system must possess the property of stability, since this is a necessary condition of its existence. However, the increase in the complexity of a living organism during evolution and development is certainly surprising since, at least from a mathematical point of view, *an increase in the complexity of a physical system generally leads to instability*. Thus, we are faced with the following questions: What is the nature of the property that allows a living organism to increase its complexity during its evolution and development, while maintaining its functional stability? What is the general principle underlying this property? And finally, can this property be used to deduce a criterion of evolution for biological systems? Before attempting to answer these important questions, let us briefly recall some aspects of the structural organization of living organisms.

Thermodynamics governs the variation of physical structures with time

The laws of thermodynamics were essentially established in the nineteenth century. Mayer (1814–1878) anticipated both Joule (1818–1899) in deriving the mechanical equivalent of heat, and Helmholtz (1821–1894) in enunciating the principle of conservation of energy. However, Carnot (1796–1832), whose pioneering work: "*On the Motive Power of Fire*" (1824), laid the foundations of the science of thermodynamics, should also receive credit for the first law of thermodynamics, which states that the total energy of a thermodynamic system remains constant although it may be transformed from one form to another. Kelvin (1824–1907) extended Carnot's fundamental ideas on thermodynamics, enabling Clausius (1822–1888) to formulate the second law of thermodynamics, whereby heat cannot pass spontaneously from a colder to a warmer body. In 1850, Clausius published the first complete theory of thermodynamics, the mathematical framework of which was later developed by Gibbs (1839–1903), Boltzmann (1844–1906) and Planck (1858–1947). In particular, Boltzmann applied the methods of classical statistics and Planck introduced the techniques of quantum statistics that revolutionized the science of thermodynamics.

The laws of thermodynamics may be usefully expressed in intuitive terms. Thus, according to the first law, *since the total energy in a system is constant, it is impossible to construct a machine that will produce work without an input of energy from an external source.* This corresponds to Carnot's affirmation concerning the impossibility of constructing a *perpetuum mobile.* And, with the formulation of the second law according to Thomson and Planck, *it is impossible to construct a continuously operating machine that does mechanical work and cools a source of heat without producing any other effects.* This implies that certain energy transformations involving heat are not reversible. For example, if there existed a process opposite to the generation of heat by friction, it would have the property of a *perpetuum mobile.* Therefore, if the second law is valid, such a process cannot exist.

Although the expression of these laws appears rather simple, it has taken a great deal of time and thought to grasp the full power of thermodynamic principles. The true significance is profound indeed. Thus, although the total energy of the universe remains constant by virtue of the principle of conservation, an increasing fraction of this energy is degraded in the form of heat that can flow only in one direction, i.e. from a warm source to a cold sink. In a closed physical system that varies with time, the quality of the energy produced or consumed therefore plays an essential role. If the system produces heat, it will never be able to return to its initial state. It will have moved irreversibly from one state to another by dissipating heat. In other words, an irreversible system is dissipative. The time-variation of such a system involves the use of the highly abstract concept of *entropy*, noted S. The variation of S, i.e. the quantity Q of the heat dissipated divided by the temperature T, describes the time-variation of the system: $\Delta S = \Delta Q/T \geqslant 0$. This corresponds to the mathematical expression of the second law of thermodynamics, which states that *in a closed system, the variation of entropy with time is positive or null, and that the entropy function is a maximum when the system is in a state of thermodynamic equilibrium.*

In other words, the state of our universe naturally tends to disorder such that its degradation is ineluctable. As Clausius put it, the energy of the universe being constant, its entropy tends to a maximum. At the macroscopic level, the homogenization of temperature in a closed physical system, the tendency towards disorder, or the progressive dissipation of useful energy into heat, are all equivalent expressions of the second law of thermodynamics. However, at the microscopic level, the increase in temperature corresponds to an increase in the number of molecular collisions. The second law thus describes a phenomenon that originates at the microscopic level and finds its expression at the macroscopic level. The passage from the local (microscopic) level to the global (macroscopic) level involves mathematics. In fact, the literal description of the phenomenon given above is built upon the mathematical concepts of spatiotemporal variation associated with the physical concepts of molecular collisions and energy.

Biological systems can change their structural organization

In an open system that exchanges energy with the environment, the variation of entropy is due to two causes. On one hand, entropy is transported across the frontiers of the system (heat exchange), and on the other, entropy is produced inside the system. According to the second law of thermodynamics, the production of entropy is positive or null. This characterizes the phenomenon of irreversibility since entropy can only be produced by means of an internal transformation. Thus, a chemical reaction, the diffusion of a dye in a solvent or the dissipation of heat from a light bulb into the ambient air, are all illustrations of irreversible processes.

As mentioned in Chapter IV, the equilibrium state of a closed system, which is an attractor for the non-equilibrium states in a certain domain of the phase space, can be determined by the Lyapunov function. For example, the non-uniform distribution of temperature in a closed system (a state of non-equilibrium) will tend irreversibly to a uniform distribution (the equilibrium state or attractor). Since the Lyapunov function and its derivative have opposite algebraic signs, the function can be used to specify the direction of the time-variation of the system. Although the method appears rather complicated, it is often useful to be able to determine whether a system is moving forward or backward in time.

Of course, the second law of thermodynamics also indicates the irreversible time-variation of a system, and clearly explains observed events, such as the diffusion of a dye in a solvent or water falling to the ground, as phenomena that vary with time from the past to the future. Why then should we go to the trouble of determining a complicated Lyapunov function when ordinary physical laws can be used to determine the sense of the time-variation of phenomena? This question will be readily answered by taking a closer look at conventional physical laws. Obviously, these laws should *not* lead to identical results when a positive value of time (for a system moving forward in time) is replaced with a negative value (for a system moving backward in time). But, in fact, they do! Surprisingly, when we replace t by $-t$, all physical laws are invariant, with the exception of the law of entropy, which may therefore be used to reveal the arrow of time.

Thus, the fundamental laws of physics appear to be contradictory. For example, if we observe the behavior of a dye diffusing in a solvent, in time the medium will irreversibly take on a uniform color. Now, let us consider what happens at the molecular level. The molecules collide according to physical laws but there will be nothing to indicate whether the collisions occur forward or backward in time. If these collisions could be filmed, it would be impossible to distinguish whether the projected film was being run forward or backward. The contradiction clearly arises from the fact that we are moving from the microscopic view of a few individualized molecules to the macroscopic view of the global behavior of the billions and billions of molecules in the system. Finally, the problem is that of the relationship between Newtonian dynamics and statistical thermodynamics, in other words, the problem of the description of irreversible processes in molecular terms, or that of *the relationship between random microscopic phenomena and irreversible macroscopic phenomena*. Prigogine, Nicolis and Babloyantz (1972) showed that the solution of this important problem, which calls for a new definition of time, has extraordinarily far-reaching ramifications in physics and biology.

Classic thermodynamics deals with systems that are in a state of stationary equilibrium, defined as a state at the asymptotic limit of thermodynamic variation where there is no change, i.e. a state without any flux of matter or production of entropy. The ordered, static structures that then appear are governed by the principles of thermodynamics. However, as we shall see below, a physical system may exist in a stationary state of non-equilibrium. Then, by producing entropy, it may reorganize itself according to a new structural order in the form of a dissipative structure. In such structures, the transport of matter is the result of the production of entropy. The fact that the conditions for the existence of such states are independent of any specific models establishes their general nature and allows a thermodynamic interpretation. If a Lyapunov function can be defined in non-equilibrium conditions, then the stability of the system will be ensured such that, after a perturbation, the system will always return to its initial state. This state is therefore an attractor for the perturbed system. Let us now take a closer look at the problem.

Prigogine, Nicolis and Babloyantz (1972) determined the conditions in the neighborhood of the equilibrium state for the second-order perturbation of entropy, i.e. a second-order difference of entropy with respect to the equilibrium state. This abstract notion is analogous to that of acceleration, which may be described as the second-order perturbation of the distance covered by a moving system. From the physical point of view, when a system is in thermodynamic equilibrium, literally nothing happens and the matter composing the system may be said to be "inert". The system may then be perturbed, for example, by mechanical agitation, or by the effect of heat or light. This perturbation, measured by a certain quantity, which is always negative, has a derivative with respect to time, which is always positive, and which corresponds to the production of entropy. We therefore have a Lyapunov function such that the displacement from the equilibrium state, which represents a small fluctuation, is stable. Lyapunov's theorem then allows us to deduce that the oscillations of such a system will always be dampened with time.

A similar result can be obtained for non-equilibrium states. If such states are perturbed, the second-order difference with respect to the state of non-equilibrium is also a Lyapunov function. However, this is true only under certain mathematical conditions[39] that may be calculated and imposed on the physical system. Thus, *away from equilibrium, the system can become stable and remain in an ordered state that differs from the state of equilibrium.* This fundamental result is illustrated by Plate V.1, which shows various *dissipative structures.* These structures require such specific conditions that only sheer chance could have led to the discovery of the Belusov-Zhabinsky non-equilibrium chemical reaction that allowed the visualization of dissipative structures. In this process, a mixture of potassium bromate, malonic or bromomalonic acid, and cerium sulfate is dissolved in sulfuric acid, and maintained away from thermodynamic equilibrium by continuous heating. With certain initial concentrations, the mixture forms a layered structure, with various shades of light and dark, showing a spatial as well as a temporal periodicity.

Thus, the phenomenon of diffusion, which generally tends to reduce differences of concentration to produce a homogeneous solution, can also lead to the formation of stable heterogeneous structures under certain conditions. Dissipative

---

[39]It can be shown that the derivative with respect to time of the second-order difference is related to the *production of entropy due to the perturbation*, and not to the total production of entropy. The mathematical sign of the derivative is therefore not determined and depends on the system.

structures of this kind are currently used as models to explain morphogenesis, i.e. the evolutionary development of form and structure in an organism. For example, certain reactions may be associated with diffusion processes, particularly when a substance playing the role of an inhibitor is supposed to diffuse more rapidly than another substance playing the role of an activator. The existence of such substances, called "morphogens", has recently been established, and Murray (1989) has shown that the mathematical solution of reaction-diffusion equations explains several of the complex patterns found in nature. Thus, the simple geometrical spiral corresponding to the solution of these equations may be transformed, by modifying the parameters of the system, into the complicated spots, stripes and other markings observed in living organisms (Plate V.2).

Of course, several other morphogenetic models may be imagined on the bases of certain physical properties, such as surface tension, phase transitions, and so on. However, we cannot expect to obtain a satisfactory explanation of the phenomenon until the morphogens involved have been identified. Nevertheless, the underlying principle appears to be sufficiently general to explain the modification of structures in a living organism. In any case, the thermodynamic theory of irreversible processes lays down the conditions under which newly organized structures may appear during its growth and development. This approach may therefore be expected to lead to the principle of evolution governing the hierarchical structural organization of a biological system.

The principle of stabilizing self-association explains the existence of functional interactions

Let us now consider the functional organization of a living organism in terms of combinations of elementary functional interactions. Since there are several ways in which functional interactions may be combined in the functional organization of the biological organism, may there not be a natural principle underlying the specific functional organization actually observed? If we assume the existence of a mechanism that allows the optimum choice of the functional organization, its mathematical formulation should predict certain observable consequences. The key to the principle of evolution of living organisms may well be found by investigating the stability of their functional organization. We may recall that, within the framework of the theory we propose (Chapter I), the stability of the biological organisms depends on the stability of their two constitutive systems, i.e. the O-FBS, representing the topology of the system, and the D-FBS, representing its dynamics. Thus, we have to examine, on one hand, the stability of the organization, and on the other, the stability of its associated dynamics. It should be borne in mind that the theoretical study presented here deals with formal biological systems, i.e. ideal systems that retain only the essential properties of real biological systems, so as to facilitate a mathematical analysis. Of course, the final objective will be to approach real biological systems as closely as possible by progressively refining the theoretical results obtained.

Observation shows that a functional hierarchy is constructed during the development of a biological organism by means of successive cell division and differentiation. In spite of the extraordinary reorganization this involves, the system maintains its stability such that the homeostasis of the existing physiological functions is conserved even as new, specialized physiological functions are incorporated into the developing system. This functional invariance is of profound significance since *the hierarchical functional organization of a biological system cannot be maintained during the emergence of structural discontinuities unless the stability of the system increases.* Obviously, if the stability decreases, the organism will not survive. In mathematical terms, the stability of a biological system must increase with its complexity. This is surely a surprising finding since it is well known that the dynamics of physical systems with increasing complexity tends towards instability. Thus, given the stability of developing organisms, we may seek to determine which of the possible functional organizations could ensure its own stability. A difficult question since it involves the determination of the essential link between the topological organization (O-FBS) and the dynamic processes (D-FBS) of the biological systems.

By analogy with the forces that stabilize physical structures in the equilibrium state, we may consider that *the physiological processes occurring during the development of a biological organism are stabilized by its functional interactions.* To illustrate this idea, let us imagine what happens during the process of cell differentiation. If each structural unit possessed all the products necessary for its own functioning, there would obviously be no need for functional interactions between the units. These units would constitute a *unique* class, containing elements identical in structure and function. Then, as we have seen in Chapter IV, this class of elements would lead to the emergence of a higher level of organization. However, if one of the units lacked some essential product, it would have to acquire this product from some other unit in order to survive. *This implies a process of self-association leading to a functional interaction between the source (the emitting unit) and the sink (the receiving unit)* (Chauvet, 1993d).

Stated in these terms, the hypothesis of self-association is a paradigm that has to be interpreted specifically in each case. For example, let us consider a structural unit corresponding to a biochemical pathway, defined as a sequence of enzymatic reactions. Then, if the biochemical pathway is modified because of the absence of a particular enzyme in one of the steps, then the missing enzyme must be obtained from some other biochemical pathway so that the modified structural unit may continue to operate normally. Here, the paradigm of self-association leads to the creation of a network of biochemical pathways, increasing the complexity of the individual pathways while stabilizing the overall processes. Of course, this paradigm will have to be expressed in totally different forms to describe complex biological and ecological phenomena, such as the physiological switch-over from glycolysis to gluconeogenesis in the fasting state, the shift in social behavior from autonomy to dependence, or from parasitism to symbiosis, and so on.

In mathematical terms, the biological system, by means of self-association between its constituent units, tends to increase its complexity to extend the domain of stability of its organization. The mechanisms behind the self-association of structural units leading to functional interactions, which are evidently of a genetic nature, correspond to a series of transformations executed according to a specific program. However, from the functional point of view, these transformations actually lead to an increase in the stability of the processes that are initiated by genetic mechanisms. Thus, we may inquire whether the biological systems of a real organism might not also be organized according to the hypothesis of self-association. The preliminary results in favor of this hypothesis (Burger, Machbub and Chauvet, 1993) have been followed up by investigations into the mathematical behavior of systems varying in time and space. As we shall see in Chapter VI, such systems correspond to a more general class of phenomena that can be interpreted by field theory at various levels of biological organization.

Finally, *the conditions for the existence of a functional interaction between two units, u_1 and u_2, are those that allow an increase in the stability of the dynamics of the system $(u_1 - u_2)$.* More generally, the conditions for the existence of a physiological function, and consequently the conditions for the emergence of a new level of organization, are those that increase the stability of the system. Thus, by observing a functional interaction and its associated dynamics, we should be able to determine the mathematical relationships between the parameters involved and obtain a better understanding of the system. Let us first consider the simple example of the functional interaction between a neuron and a skeletal muscle fiber. This consists of the action of an electric potential on the muscle fiber, which then undergoes a contraction. Now, a skeletal muscle fiber that ceases to receive nerve signals will die. In other words, neuronal stimulation is the essential condition of the survival of the muscle fiber and the product of the functional interaction it receives ensures its stability. This is a striking illustration of the principle of stabilizing self-association. The mathematical conditions for the stability of the system may

V

then be used to investigate the functioning of the specialized neuromuscular junction known as the *motor endplate*. Let us now consider the somewhat more complex example of the network of metabolic pathways. This example is of considerable interest since it has allowed us to describe a whole class of dynamic systems and establish the paradigm of self-associative stabilization.

Networks of metabolic pathways are created by self-association

Let us take another look at Figure IV.5, which represents a hierarchical metabolic system operating at two levels, i.e. the epigenetic and the metabolic levels. We have already considered a structural unit corresponding to the metabolic pathway for the synthesis of a product in a certain number of enzymatic steps (Figure II.1). From the biochemical point of view, the typical process is described by the epigenetic system shown in Figure II.3. This constitutes a structural unit, in which each of the required enzymes belongs to a hierarchical functional system, corresponding to the biosynthesis of proteins. The fundamental level, or level-0, consists of the structural units represented, on one hand, by the regulatory genes, the structural genes and the polysomes, and on the other, by the functional interaction represented by the messenger RNA, such that the global dynamics at level-0 produces the enzyme. This enzyme then enters a biochemical pathway at level-1, the global dynamics of which leads to the final product. Since the time scales at the two levels are very different, the variables of state at level-0 can be considered as constants, i.e. parameters, at level-1 (Chapter II). This is of course only a schematic description of a metabolic system and we may be sure that a real biological system would be infinitely more complicated. Fortunately, Goodwin (1976) has established the mathematical conditions of validity that are broad enough to allow considerable simplification of metabolic systems.

Although the elementary representation of the metabolic system given above describes the essential mechanisms involved, many problems will have to be solved before we obtain a satisfactory insight into the metabolic process. For example, let us imagine the chemical reactions described in Chapter II being carried out in a cell, or in some subcellular organelle, such as the nucleus or a mitochondrion. There may be hundreds of enzymatic reactions occurring simultaneously in thousands of metabolic pathways, but each particular enzyme has to act on a specific site. But how do we explain the transport of a given enzyme from its point of origin in a metabolic pathway to its specific site of activity? It has been suggested that the metabolite in a biochemical pathway is carried from one step to another by a "channeling" mechanism allowing it to move in the surrounding medium as if in a space limited by a structural discontinuity. Such channels would thus correspond to "molecular routes" ensuring the direct transfer of metabolites to their destination.

Unfortunately, it is very difficult to prove the existence of these molecular routes. The kinetic isolation may be created by a diffusion barrier, or by means of a distance greater than that of the other cellular processes, or even by the segregation of a metabolic pool within a specific micro-environment, as in the cell membrane. But how are the various channels corresponding to specific metabolic pathways actually separated to avoid interference? With so many questions unanswered, the channeling concept has raised much controversy concerning the significance of the experimental results obtained (Ovàdi, 1991).

An interesting functional approach to the problem, proposed by Westerhoff and Welch (1992), is based on the biophysical organization of molecular structures. This involves evaluating the rates of transfer along a metabolic pathway to determine whether the transport is due to a process of physically limited diffusion in the neighborhood of the metabolic pathway rather than to a phenomenon of general diffusion in the medium. Our own approach, as presented above, is also of a functional nature and may be expected to lead to an understanding of the organization of complex networks of metabolic pathways.

Thus, the investigation of stability in the neighborhood of a stationary state shows how the concentrations of the intermediate products in the hypothetical "channel", as well as in the body of the solution vary with the rate constants of transport. It has been demonstrated that the stability, in terms of the rate of convergence towards the stationary state, increases when the metabolic pathway is confined to a reduced volume, such as that of a cellular substructure, and when metabolites can be exchanged between this restricted space and some other space, such as the cytosol (Chauvet and Costalat, 1995). The channeling mechanism thus appears to have a stabilizing effect in comparison to the simple sharing of metabolites. Furthermore, when the metabolic pathway stretches across two distinct spaces, the exchange of metabolites between the two portions can be shown to increase the stability of the whole system.

The stability of the dynamics at the molecular level allows us to consider a metabolic pathway as a *structural unit*, say u_1. This unit consists of a set of molecular enzymes and substrates that are *channeled* and *stabilized* when the exchanges occurring between the locally structured space and some other space involve not only the passive phenomenon of diffusion but also electrostatic effects and differences of affinity between the metabolites. This suggests the existence of a structural discontinuity since the diffusion considered here is limited to a certain space. In effect, the phenomenon of diffusion corresponds to a purely thermodynamic process of homogenization, which could not otherwise be restricted to the neighborhood of a metabolic pathway. We may wonder whether a structural unit such as u_1 would tend to associate with another unit to form a network of metabolic pathways. If this were the case, it would imply the existence of functional interactions between the two metabolic pathways. This would give rise to a new structural unit: $u_2 \equiv (u_1 - u_1)$, by the association of two units of the type u_1, separated by a structural discontinuity. This new unit could then lead to a higher level of organization, i.e. a network of metabolic pathways.

This example of the association of metabolic pathways in a network allows us to interpret the paradigm of self-association mentioned above. Let us suppose that the metabolic unit u_1 undergoes some modification, for example, following a micro-mutation that blocks the synthesis of an enzyme E_2 (Figure V.1), such that the metabolic pathway is broken just where this enzyme would be required. Then, according to the hypothesis of self-association, the survival of the modified unit would require that it enter into association with a normal unit. This would lead to the formation of a new structural unit at a higher level of complexity. The stability of this new structure may be investigated by studying the dynamic properties of a system comprising two identical metabolic pathways (Figure V.1).

As we may recall, Figure IV.5 represents the case of a single metabolic pathway. In this schema, the dynamics of a system (u_1) are represented by changes in the four state variables cor-responding to the succession of products elaborated in the pathway. The terminal product has a retroactive effect on the concentration of the first enzyme of the pathway through an inhibitory interaction of a cooperative nature. In other words, the binding of a ligand to a given site on the molecule reduces the affinity of the ligand for all the other sites of the molecule (Chapter II). The mathematical structure of the dynamic system representing these kinetic mechanisms leads to a classic study of the stability of solutions, whereby the parameters of the chemical reactions, i.e. the rate constants, *must* satisfy the cal-culated conditions. If this were not the case, the chemical reac-tion would be uncontrollable and the concentration of the final product would tend to infinity rather than to that of the sta-tionary state. After the association of two units, such that $u_2 \equiv (u_1 - u_1)$, the dynamics of the new system (u_2), compris-ing two metabolic pathways, is described by five state vari-ables instead of four (Figure V.1). The stability of this system, investigated by the same methods as for the single metabolic pathway, gives the results shown in Figure V.2.

And here again we find the surprising result we have already mentioned. *In spite of the greater complexity of the new system (u_2), comprising two metabolic pathways, its region of dynamic*

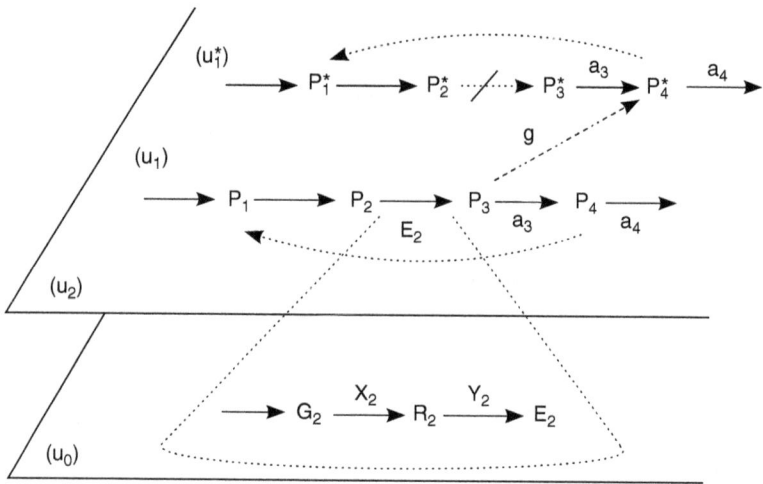

Figure V.1: The two structural units, u_1 and $u_1{}^*$, are identical except that $u_1{}^*$ has undergone a modification: the chemical reaction from P_2 to P_3 no longer exists so that the other reactions of the pathway are suppressed. This biochemical pathway will disappear unless the molecule P_3, which is at the origin of the missing links (such as $P_4{}^*$), is supplied by u_1. We then have a set of two metabolic pathways, associated through the metabolite P_3 with a coupling factor g. The α values are the rate constants of the chemical reactions. The enzyme E_2, which leads to the elaboration of P_3, is produced, like the other enzymes, by the epigenetic system u_0 situated at a lower level. The units, u_1 and u_0, are the same as those in Figure IV.5.

stability is greater than that of the system (u_1), *which has only one pathway.* Thus, the creation of a functional interaction between two associated metabolic pathways actually has a stabilizing effect. This is a powerful argument in favor of the evolutionary development of networks of metabolic pathways in biological organisms. Among the numerous pathways by which a product might develop, the only observable pathway will be the one that corresponds to an interaction with another pathway since only then will the system tend to a stationary stable state. Although, we have investigated a network comprising only two metabolic pathways, the dynamics of more complex networks, based on the same mathematical structure, may be expected to yield similar results. Since the

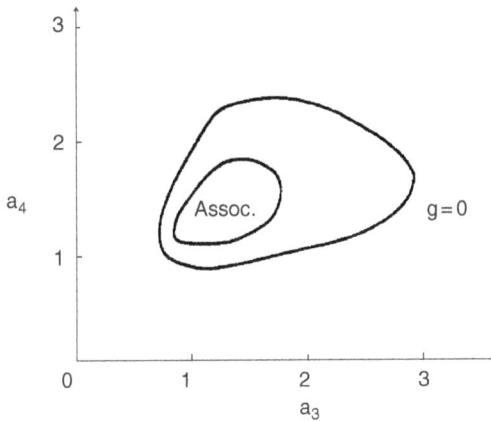

Figure V.2: The areas lying outside the closed curves represent regions of stability. The regions shown correspond to the network comprising the two metabolic pathways, u_1 and u_1^* (Figure V.1), associated by a coupling factor g. The dynamics of the system are represented in the phase space of the system, i.e. the space of the two rate constants of reaction: a_4 and a_3, for two values of the coupling factor g, one being null and the other non-null. The calculated region of stability increases when g changes from a null value, corresponding to the absence of association between the two pathways, to a non-null value, corresponding to an association between these pathways. In other words, the association between the two metabolic pathways leads to an increase in stability of the system.

equations obtained are quite general and contain terms taking into account the phenomena of reactions and transport, they should prove useful in a wide variety of biological systems.

Finally, in biology, as in the other sciences of nature, we have to strive to explain what we observe in living organisms, i.e. find a logical interpretation for observed facts. Why, for instance, do various types of cells have to associate with each other to fulfill their functions? Why, among several possible organizations, do we find a particular network of biochemical pathways rather than another? Why does a metabolic pathway need to be circumscribed? *The simple answer to these questions is that the organization of the interactions between groups of cells in the body, or between the metabolites in a biochemical pathway, is the end result of a process of self-association that takes place during the development of a living organism so as to increase the stability of each of its dynamic systems.* The problem we then have to solve consists in determining the conditions of stability of biological systems and establishing the consequences for the living organism. However, rather than the usual approach based on the study of the stability of some particular process, what we propose here is a more general investigation of an integrated biological system comprising its structural units and the corresponding functional interactions. As we shall see in the next chapter, this important problem can be tackled within the framework of an appropriate field theory.

A novel qualitative model describes the specialization of tissues in living organisms

During the development of a biological organism, a tissue initially contains a large number of cells, all apparently having the same function, for example, each of the cells may secrete a specific molecule destined to act on some other cell. In fact, the genetic program of the organism coordinates the complementarity between the structural units. Thus, the development of the tissue involves the proliferation of specialized cells under certain physiological constraints. If these physiological constraints fail to play their role, pathological situations, such as cancer, are likely to arise.

According to the hypothesis of self-association, a unit that undergoes a modification, such as a micromutation, must, in order to survive, associate with a normal unit so as to form a new structure at a higher level of organization, which in turn will give rise to a new population, and so on, until the organism is fully developed. Successive self-associations lead to the creation of a new level of organization corresponding to a new physiological function. This process, which is closely analogous to that of tissue specialization, has the same biological significance as organogenesis, except that here the phenomenon is governed by the programmed expression of genes rather than by micromutation. Without going into the details of the mechanisms involved, we may simulate the evolution of a population of units, at least from the *qualitative* point of view, by means of simple reasoning based on the principle of successive self association. The lower part of Figure V.3 corresponds to the model of the first few segmentations of a fertilized egg cell as shown in Plate V.3. The embryonic transformations may be interpreted as follows: at the beginning of the transformations, i.e. at the initial instant of the model, the two products, P_1 and P_2, are assumed to be missing in a certain unit u_1^*. To synthesize these products, u_1^* will associate with another unit, u_1, to form a new unit, u_2. A series of transformations of this type will lead to the stage represented in Figure V.3, in which the structural unit u_4 is specialized in the elaboration of another product, noted P_3.

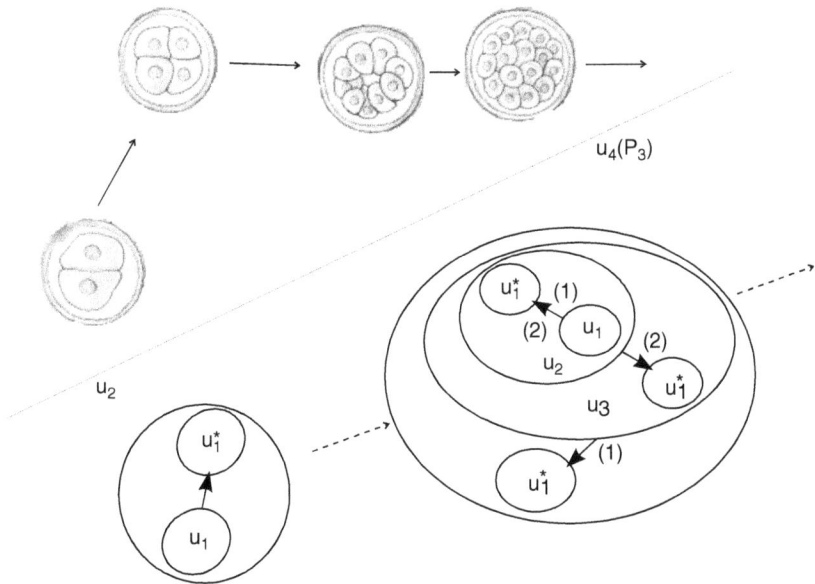

Figure V.3: A qualitative model of the process of segmentation during organogenesis. The upper part of the figure shows the first segmentation steps starting from the initial cell. The lower part shows how segmentation may occur by self-association according to the model proposed. The products exchanged between structures at different steps correspond to the creation of functional interactions. The hypothesis of self-association applies to the successive cell divisions and differentiations during embryonic development as illustrated in Plate V.3.

V

A great deal of work has been done on the phenomena of morphogenesis and cell differentiation. It is widely believed that the knowledge a cell acquires of its environment or, more precisely, of the surrounding cells, allows it to synchronize its own development with that of its neighbors. The existence of morphogens, i.e. the molecules that were hypothesized to explain how cells recognize each other, has recently been confirmed. However, the behavior of a cell with respect to its neighbors could also be explained by *the application of functional constraints to the mechanisms involved in the creation of structures during the development of a biological organism.* In other words, the specialization of tissues formed by cell differentiation may be considered as the result of a series of functional self-associations leading to a higher level of complexity. Thus, a genetic program would allow a transformation, for example $u_1 \rightarrow u_1^*$, by the modification of some enzymatic reaction such that the increase in stability necessary for the system would initiate a process of self-association. As we shall see below, the phenomenon of specialization decreases the "functional complexity" of a biological system while increasing its "functional order". Moreover, we shall demonstrate that the principle of biological evolution is an optimum principle and that the formal biological system can evolve only by increasing its functional order through a process of specialization.

Physiological invariance is governed by the principle of vital coherence

The increase in the stability of a system in spite of an increase in its complexity is, as mentioned above, an essential characteristic of living organisms. This suggests the existence of a general *principle of vital coherence* that ensures the invariance of the physiological functions of a biological system by the self-association of its structural units.

The structural units of an organism occupy subspaces limited by structural discontinuities in physical space. The mechanism of self-association transforms the structural units into sources

and sinks by the introduction of functional source-to-sink interactions, i.e. the functional organization of the system consists in
the concomitant creation of sources and sinks. If structural discontinuities did not exist, the transport of matter between the
structural units would occur only by *free diffusion* through
simple thermodynamic effects. In effect, this is the way molecules are transported within the compartments formed by the
structural discontinuities at the lowest level of the structural hierarchy. The mechanism of transport is easy to understand in the
case of blood circulation when a molecule carried by the blood
from one part of the body to another by convection finally dif-
fuses within the cytoplasm of the target cell. However, the
mechanism of transport may be far more complex in the case of
transmembrane transport requiring transformations within the
membrane, as for example when ionic pumps are involved.

The principle of vital coherence establishes the *existence* of
a biological system represented by a set of functional, dynamic
interactions between its structural units. In this sense, it generalizes the hypothesis of self-association, according to which a
structural unit that loses an elementary, indispensable physiological function must retrieve this physiological function from
some other structural unit in order to survive. It also corresponds to the phenomenon of *adaptation* in evolutionary science. Since self-association increases the number of structural
units in an organism, the *degree of organization* is defined as
the number of structural units at a given level of a biological
system. However, does the application of the principle of vital
coherence lead to an increase in the stability of the system?
As we have seen above, this is probably true at the lowest
level of organization, i.e. the level at which molecular transport
occurs by diffusion. Although this question is far more difficult
to tackle at the higher levels of organization, we shall see that
the application of a field theory (Chapter VI) offers a promising approach.

An important aspect of the hypothesis of self-association is
that it incorporates the *conservation* of the physiological function, which is the global result of a set of functional interactions in a biological system. For example, muscle contraction

is conserved when a neuron is associated with a skeletal muscle fiber, just as nervous activity is conserved when one neuron is associated with another. The principle of vital coherence is in fact the *principle of conservation* of elementary physiological functions, i.e. of the products necessary for survival. Expressed in this form, the principle of vital coherence allows us to define *functional organization* as the distribution of the number of sinks according to the functional interactions involved, thus giving a quantitative representation of the topology of the biological system.[40] *We may then define the general functional organization of the system by taking into account the individual products that may be missing.*

Similarly, in animal societies at the highest level of organization, the problem of survival appears to have been solved by the association of individuals in large groups. In mathematical terms, the phenomenon of association is necessarily related to the stability of the dynamic system, which cannot but tend towards its attractor, i.e. its state of equilibrium. This would explain how animal societies are formed and how symbiotic phenomena appear.

Is the principle of vital coherence, which is of a topological nature, sufficient to explain the hierarchical functional organization of a biological system? Clearly, certain conditions will have to be imposed on the system so that a new physiological function emerges. One such condition, discussed below, is that of *specialization*. However, since we have defined a physiological function in terms of the global dynamics of a set of units of a formal biological system (D-FBS), other conditions of a dynamic nature will be required if the set of functional interactions is to constitute a new physiological function. These conditions, related to *specific time scales*, will be discussed in Chapter VI. The association between units during the development of a biological organism leads to the emergence of a well-defined physiological function operating on a specific time scale. For example, at the cellular level, the association

---

[40]The functional organization may therefore be represented either as the *topology*, the *graph*, or the *distribution of sinks* in the system.

between neurons produces the functions of memory and learning. Similarly, at the organ level, the association between the hypothalamus, the hypophysis and the thyroid leads to the homeostasis of the metabolism of the organism.

The principle of vital coherence, which corresponds to the natural principle of invariance, a phenomenon that appears in all biological systems, is by no means tautological since it expresses not only the remarkable physiological invariance but also the extraordinary adaptation of the functional organization of living organisms to various situations. These situations are of two types: physiological and pathological. Physiological situations are "predictable", as in the case of genetic transpositions, and during successive steps in cell differentiation or morphogenesis. In contrast, pathological situations are "unpredictable", as in the case of mutations or structural alterations of the genome.

As we have seen above, the principle of vital coherence is a principle of conservation. However, from a mathematical point of view, it should be observed that it is not the *limits* of the *value* taken by a given physiological function that are conserved, but rather the *existence* of the physiological function itself or, in other words, the *existence* of the corresponding dynamic system. Let us now take a closer look at these two important aspects of physiological invariance, i.e. the *value* and the *existence*, of physiological functions.

Physiological invariance determines the value and existence of physiological functions

The *value* of a physiological function — the first property of physiological invariance — is a consequence of the *principle of homeostasis*. We may recall that a physiological function may be mathematically represented by the solution of a dynamic system, i.e. the D-FBS. The solution of the system, which corresponds to the global behavior of a set of structural units, depends on the state variables that describe the time-variation of the system. As we shall see in Chapter VI, the dynamics of physiological processes can be conveniently represented by means of a field theory.

V

Let us consider a physiological function, such as blood flow or blood pressure, in terms of a mathematical function containing several state variables. It is a biological *fact* that the quantities represented by these variables possess the intrinsic property of "variability" due to random fluctuations according to the laws of probability. If a given quantity varies according to a certain law of probability, then we can calculate the chances of finding its value within a determined interval, and thus study its average variation with time. The mechanisms of control of such a system are formally represented by the relationships between the quantities involved, and the effect of the control is such that the values of the corresponding variables are maintained within certain limits. If these relationships describe local mechanisms, i.e. mechanisms that vary within an infinitely small time interval, then we obtain a set of differential equations with respect to time, the solution of which constitutes the *dynamics* of the system.

As we have mentioned above, the survival of a biological system requires that its dynamics be stable. In other words, the associated biological quantities must have values that tend towards those of the stationary state while being maintained within certain limits by the regulatory mechanism. For example, if the biological quantity corresponding to blood pressure tends to zero in a patient under intensive care, the organism must be in a zone of physiological instability. Obviously, since we are dealing with a biological system, all the other quantities associated with the blood pressure will also have varied such that they may no longer be within their intervals of stability. These observations may be graphically represented as a set of points moving in a space of many dimensions, passing from one region to another, with critical boundaries between regions of stability and instability. In Chapter VI, we shall see how the stability of the dynamic system of functional interactions, i.e. the stability of the D-FBS, can then be determined by means of a suitable field theory.

The *existence* of the physiological function — the second property of physiological invariance — is a consequence of the *principle of vital coherence*. To illustrate this property, let us consider a case in which the pancreas ceases to function, so that the lack of insulin leads to diabetes. This dysfunction may be due to several causes, such as poor organ regulation, degeneration or a tumor. Thus, in the topological representation, i.e. the O-FBS, representing the regulation of blood sugar, the pancreas will be missing. It is easy to imagine the pathological consequences of the absence of the pancreas, but how are we to describe the suppression of an organ in a manner that is sufficiently general to include *all* the observed facts belonging to the same category, i.e. the suppression of an element of the organization of the system? In a very general and abstract way, the property of *existence* may be represented by the topological stability of the functional organization, i.e. the stability of the graph representing the O-FBS. In effect, we have to determine the consequences of the suppression of an organ at the summit of the graph. Suppressing the summit of a graph means the suppression of the product emitted by the summit, considered as a source, i.e. the modification of the summits corresponding to the sinks for this product. If the physiological function corresponding to the global behavior of the graph is to be conserved, it will be necessary to reorganize the graph by *modifying the distribution of the interactions between the summits.*

Unfortunately, in the case of the pancreas, there is no substitute for the organ so that when the pancreas fails, its physiological function will not be conserved. However, it is noteworthy that insulin, secreted by the pancreas, is the only hormone that reduces blood sugar whereas the body produces no less than five hormones with the opposite effect, i.e. glucagon, adrenalin, cortisol, cortisone and corticosterone. What could explain the abundance of sites producing hyperglycemic hormones compared to the unique site producing a hypoglycemic hormone, all these hormones participating in the same physiological function, i.e. the regulation of blood sugar? From a general point of view, we have to determine the mathematical conditions such that the property of invariance is satisfied, since pancreatic failure reflects the absence of invariance. In fact, the knowledge of the conditions necessary for the emergence of invariance may be expected to lead to an insight into the true organization of the physiological functions of the organism.

Let us now suppose that all biological systems faced with constraints reorganize themselves to ensure the permanence of their functions. This may be considered to represent the principle of vital coherence in its *restricted* form. In contrast, the *general* form of the principle would be considered applicable when only certain physiological functions are conserved. As a conservative principle, it ensures that the functional interactions reorganize themselves by association such that the global activity existing before the reorganization subsists after the process. In other words, *the physiological function remains invariant during the successive transformations of the functional organization of a biological system.* The restricted form of the principle of vital coherence expresses the conservation of the physiological functions *essential* to the life of the organism. Like other principles, it may appear tautological since it is valid in a large number of situations. However, this principle is based on the concept of the source, the sink and the functional interaction, on one hand, and as we shall see, on the concept of the hierarchical functional system, on the other.

Once the concepts and the methodological framework have been defined, certain questions immediately arise. For example, how does the functional organization of an organism satisfying the principle of invariance behave when it is subjected to internal or external perturbations? Why do we observe a particular organization among the many that may be possible? And so on. Such questions lead to two fundamental problems:

(1) How does a biological system grow and reproduce, restructuring itself by increasing the number of levels of organization and functional relationships, while conserving, on one hand, the stability of its functional organization, i.e. the O-FBS, and on the other, the dynamics of its functional interactions, i.e. the D-FBS?

(2) Is the time-variation of the functional organization of the biological system governed by an extremum principle? In other

words, is there a functional equivalent of the second law of thermodynamics that might apply to living organisms?

These are of course difficult problems to tackle. However, in view of their interest, we shall try to approach the solution within the context of the search for the general principles of theoretical physiology. Let us now consider some aspects of the nature of the functional organization of a biological system, i.e. the O-FBS.

The potential of functional organization taps a "reservoir" of developmental possibilities

Since each structural unit of the functional organization of a biological system has a fundamental gene level, it is capable of *self-reproduction*. And since all the structural units contain identical genetic material, they have the fundamental property of *equipotentiality*. In other words, each structural unit normally possesses a potential of self-reproduction identical to that of all the others. Although equipotentiality is a general biological property, the mechanism of differentiation leads to the specific development of the structural units of an organism during its growth. As we have already seen, the hypothesis of *self-association* implies that structural units must be either sources or sinks for a given product, with functional interactions operating between sources and sinks.

It is obvious that these two properties of self-reproduction and self-association play an important role in the evolution of a biological system since they signify that the functional interactions between the units may occur in several different ways between a subset of sources and a subset of sinks during the elaboration of a given product. We may then introduce the idea of the *potential of functional organization* to explain the fact that among all the structural units that could interact among themselves, since they all possess the same potentiality, only a certain number actually interact while the others remain, as it were, in reserve. This suggests the existence of a reservoir of possibilities of organization — an *organizational reservoir* — in which the system could draw during its evolution. This approach to a fundamental biological problem is similar to that used in physics with regard to the principles of least action (Chapter I). Thus, from a physical point of view, the potential energy of a body falling under the action of a gravitational force will be transformed into kinetic energy during its movement. The potential energy of the body constitutes an "energetic reservoir" from which energy is released during the movement of the body. This "energetic reservoir" of a physical system is analogous to the organizational reservoir of a biological system.

The existence of such a potential of functional organization would explain how a biological system, subjected to a series of perturbations, e.g. during the process of cell differentiation, might transform its *potential* organization into a series of *effective* organizational processes. Plate V.4 illustrates the process of hematopoiesis, i.e. the formation of platelets, red blood cells, polynuclear cells, monocytes and lymphocytes, which represent the final forms of cells derived from the various cell lines issuing from a common stem cell in the bone marrow. Thus, the *potential* organization appears to lie in the common stem cell and the *effective* organizations are manifested in the figured elements of the blood. Although the concept of the potential of organization is of an abstract nature, it clearly expresses the adaptive reality of the genome since all nucleated cells have identical potentials. We may therefore consider that the topology of the biological system varies during growth because of its intrinsic potential of organization. For example, an undifferentiated tissue, in which all the cells have an identical individual potential of organization, will be transformed at a given instant of its development into a differentiated tissue, in which certain cells will elaborate all or some of their individual physiological products. In short, since each unit must possess all its individual functions in order to survive, the graph of the functional organization must reveal the functional coupling between units necessary for the life of the system.

The *potential of functional organization* may be mathematically determined according to the properties it is expected to satisfy. These properties are suggested by our present knowledge of general physiology. First, the potential of functional organization, called Π, should be able to indicate the number of potential functional interactions of a biological system. Secondly, it should allow us to describe the functional hierarchy of the system by taking into account the different levels of organization; thus, the value of Π should vary directly according to the number of units at a given level,[41] thereby defining the hierarchy of the system. Finally, when calculated for the whole system, taking into account all the levels of organization, the value of Π should increase with the number of levels as the organism develops, but decrease when the hierarchical system undergoes reorganization. Biologists have long been familiar with these properties and, in particular, it is well established that the degree of specialization increases with the level of organization of a biological system.

As we may recall, in a network of hierarchical metabolic pathways, one biochemical pathway replaces another when a functional association is established between one set of units that possesses a certain function and another set that does not. We may imagine that this association corresponds to one of the possibilities contained in a "reservoir" of possibilities representing the potential of functional organization of the system. Thus, among the set of organizational possibilities, on one hand, and the set of biochemical pathways, on the other, the evolution and the development of the organism must have led to the choice that we actually observe, i.e. the switch-over from one particular biochemical pathway to another. This is precisely what is observed, for example, in the fasting state, during which the biological system switches over from the glycolytic pathway to that of gluconeogenesis (Chapter II). Then we obviously have to address the question as to *why* this particular choice was made rather than some other among all those that were possible.

---

[41]Since the number of structural units generally decreases as the level of organization of a biological system increases, the value of the potential of functional organization should also decrease.

Of course, the precise representation of the functional organization of a biological system in terms of its functional interactions is bound to be very difficult. However, just for the sake of argument, let us suppose this has actually been done. Now, we know that a functional interaction between two identical units increases the stability of the D-FBS constituted by these units. We also know that there is a close relationship between the functional hierarchy of the O-FBS and the structural hierarchy of the system (see Figure IV.4). Finally, we know that the

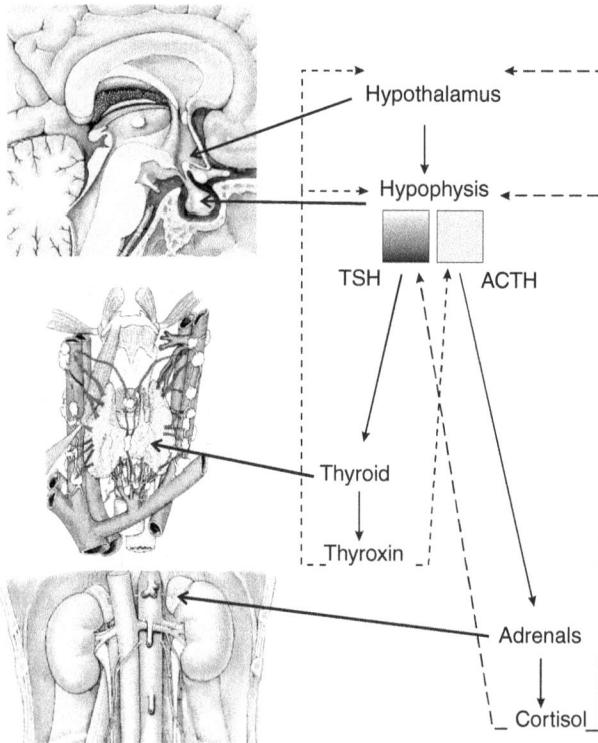

Figure V.4: Coupling and regulation of the hypothalamo-hypophyso-thyroid system and the hypothalamo-hypophyso-adrenal system. The former secretes thyroxin while the latter secretes cortisol. The anatomical localization, i.e. the structural organization, is shown on the left of the figure. The schema on the right shows the coupling between the two hormonal systems. TSH = thyroid stimulating hormone; ACTH = adrenocorticotropic hormone.

O-FBS has a potential of functional organization that can be calculated from the interactions between the units of the system. Now, let us consider two examples, emphasizing respectively the functional and the structural hierarchies of a system.

The first example is that of the hormonal system. Hormonal systems are generally classified into three main categories according to the number of regulatory feedback loops involved. For instance, the brain acts through the hypothalamus on the hypophysis, which in turn acts on some specialized endocrine gland. Certain neurohormones secreted by the hypothalamus specifically increase or decrease the secretion of hypophyseal hormones. These hormones act on "peripheral" endocrine glands, each gland secreting its own "peripheral" hormones to produce specific effects on the target cells of the organism.

Figure V.4 illustrates two hormonal systems: the adrenal system, which secretes cortisol, one of the hormones that has an effect almost the opposite to that of insulin; and the thyroid system, which produces thyroxin, a hormone whose chief function is to increase the rate of cell metabolism. Now, let us attempt to extract the elements describing the organization underlying this complex phenomenology. In fact, what we have here is mainly a functional organization, although each unit of the system is of course a structural hierarchical system with its own tissues, cells, organelles, and so on. Moreover, there are non-symmetrical interactions between one organ and another, with feedback loops operating at a distance after local transformations in each of the organs.

But how is the functional hierarchy established? First, the structural units associated with a given function must be related by interactions operating on the same time scale. These criteria define the functional level, for example, that of the secretion of cortisol or the production of thyroxin. However, as Figure V.4 shows, these two hierarchical endocrine subsystems interact, each subsystem being subjected, at a given level, to the action of the other, at another level. Admitting the hypothesis of self-association, the increase in complexity by successive self-association during development, which led to the functional association of these two hormonal subsystems, must have been accompanied by an increase in the stability of the organism. As we may recall, the possibility of such an increase in stability following an increase in complexity was discussed above in the context of the principle of vital coherence. The interactions between the two subsystems produce a system that is very similar to the metabolic system shown in Figure V.1. Thus, the mathematical formulation of the dynamics of these subsystems follows the same lines as that of the network of metabolic pathways considered above. However, although the calculations here are bound to be rather difficult, the main problem lies elsewhere since we have to take into account *all* the structural units that must have led to the emergence of these functions and therefore must belong to the same hierarchical level of the functional organization.

The second example is that of the respiratory system, some mechanisms of which were discussed in Chapter III. This is far more complicated than the hormonal system described above because it is situated at a very high level of complexity in the organism. However, it is of great interest from the structural point of view since the description of the associated physiological functions requires a variety of formalisms related to mechanics, as in the study of the bronchial constraints on air flow or to chemistry, as in the analysis of the chemical kinetics of the exchanges of oxygen and carbon dioxide in red blood cells.

In the case of respiration, from a purely intuitive point of view, we may say that two structural units — represented by the lungs — are associated with a set of neurons, which form the controlling unit, and a set of muscles that mechanically vary the intrathoracic pressure so as to perform the ventilatory function. Ventilation is therefore at the highest level of functional organization considered here. Each of these units, i.e. the lungs, the controller, the thoracic and abdominal muscles, constitutes a hierarchical system for which we may determine the level of structural organization. In the structural unit corresponding to the lungs, at the immediately lower level we have the large bronchi and the finer bronchioli, i.e. the structural units operating in association to ensure the flow of air in the respiratory system. According to the usual representation of the bronchial tree (Weibel, 1963), the bronchi subdivide in binary fashion such that, if the trachea is considered as the first generation, the number of bronchi or bronchioli at the ith generation is 2^{i-1}. The bronchi and the bronchioli from the 1st to the 16th generation interact functionally with each other, as well as with the muscles, according to the laws of fluid mechanics and continuous elastic media to ensure the flow of air. The histological structure of the respiratory system varies considerably along the airways, from the mouth down to the alveoli, but as far as our formal representation is concerned, this corresponds to a continuous variation in the elasticity of the walls of the tube-like structures. The physiological function corresponding to the flow of air, measured by the concentration of oxygen or carbon dioxide at different points of the bronchial tree, has two mechanical components, i.e. the difference of pressure acting on the air at the extremities of the tree, and the force acting on the elastic walls. Mathematically, this can be calculated by means of the Navier–Stokes equations of fluid mechanics. The alveoli, which can be considered as particular cases of the bronchioli from the 16th to the 23rd generations, interact with the capillaries to allow gaseous exchanges by diffusion; the bronchial tubes and the alveoli being associated at a single level of organization (Chauvet, 1976; Chauvet, 1978a). In the dynamics described by the Navier–Stokes equations for the flow of air in the bronchi and by the equation of diffusion for the gaseous exchanges in the alveoli, these units are taken to be identical. Thus, a bronchus or an alveolus is considered to belong to a class of structural equivalence containing all the bronchi and alveoli. If this were not so, then we would have to consider several different classes corresponding to the various anatomical characteristics of the structures. Of course, in the last analysis, each bronchial tube and alveolus is a specific structure and we would end up with as many equations in the system as there are structural units, making the problem almost inextricable. This difficulty can be resolved by considering a limited number of structural units, each representing a particular region of the lung, with functional interactions between these units (Chauvet, 1978b).

In the case of a globally regulated system, we see that the potential of functional organization depends on the number of functional interactions that may be obtained by applying the hypothesis of self-association in accordance with the principle of vital coherence. Clearly, the higher the level of the structural and functional hierarchies, the smaller is the number of structural units. Thus, in the respiratory system, the number of cells is greater than the number of air passages, which in turn is greater than the number of lungs. As the level rises, the number of potential self-associations decreases, as does the potential of organization. However, since the number of potential self-associations due to interactions between the different levels increases, the overall potential of organization of the system, taking into account all its levels of organization, actually increases. Again, we may consider a given functional system as part of a larger hierarchical system in the organism. For example, the physiological function corresponding to the maintenance of the acid-base equilibrium in the body involves the lungs as well as the kidneys. Here, two different types of unit at the same level of structural organization, i.e. the lungs and the kidneys, constitute two parallel hierarchical systems, each being at a different level of functional organization. This schema corresponds to the '*relativity*' of the hierarchical systems of biological organisms described in Chapter IV, and could be easily incorporated into the metaphorical Louvre Pyramid shown in Figure IV.3.

The real hierarchical systems presented above are necessarily incomplete, since all the mechanisms involved have not yet been elucidated. Given the complexity of real systems, we are obliged to resort to some simplification if we wish to extract the essentials without denaturing the reality of living organisms. Of course, the aim of all theoretical work is to organize the knowledge acquired in a given field, but a theory should always be open to modification when new knowledge is available. In our top-down approach, we have chosen to investigate a particular class of formal systems, the analysis of which is expected to lead to the properties actually observed in real biological systems. Let us remember that scientific endeavor, based on this kind of to-and-fro between theory and observation, has always been stimulated by the cross-fertilization of ideas from various disciplines.

Can the potential of functional organization of a biological system be calculated?

The complexity of real biological systems makes the calculation of the potential of functional organization a forbidding task. A mathematical function offering a coherent description of the functional organization of a biological system would certainly be useful. We have therefore sought a function corresponding to the non-symmetrical functional interaction between a source and a sink, as postulated in the hypothesis of the self-association of structural units during the development of a living organism. Thus, if a biological system has v structural units, among which there are n sinks, each of which may be associated with one of $(v - n)$ sources, then the number of associations[42] is given by: $o = \Pi(v - n)^n$. However, as we shall see below, it is preferable to consider the sum of terms, each of which is the product of the number of sinks and the *logarithm* of the number of sources, at each level of organization and for each functional interaction. Thus, the expression for the potential of functional organization[43] may be written:

$$v = \Sigma \ n \ln (v - n), \quad \text{or} \quad \Pi = \Sigma \ [sinks] \ln \ [sources].$$

This mathematical function has two essential properties. First, *the potential Π is non-symmetrical,* i.e. if we interchange n, the number of sinks, and $(v - n)$, the number of sources, the value

---

[42]From the mathematical point of view, this is the product, at a given level of organization, of the number of injections of the set of sinks into the set of sources.

[43]This expression may be written in the form: $\Pi = \Sigma\Sigma n_\alpha{}^i \ln(v^i - n_\alpha{}^i)$ where v^i is the total number of structural units at level i and $n_\alpha{}^i$ is the number of sinks at this level for a given product $\alpha = 1, \mu$.

of the function is altered. We may recall that non-symmetry is a fundamental characteristic of biological laws (Chapter I). Secondly, *the potential* Π *has a maximum value for a certain number of sinks* (Figure V.5). This means there must exist a specific organization, i.e. a particular distribution of sinks involved in the functional interactions of the system, such that the potential of functional organization has a maximum value.

This expression of Π suggests two extreme possibilities: one in which no sinks exist $(n = 0)$, and the other in which there is just one source $(v - n = 1)$. If we transpose this idea to an animal society, the former would correspond to a completely independent individual, ensuring the elaboration of all the products required, and therefore needing no units corresponding to sources. In contrast, the latter possibility would correspond to a highly specialized individual supplying a certain product needed by all the other members of the society. In each of these two cases, which are of course the simplest we can imagine, the potential of functional organization is null. If,

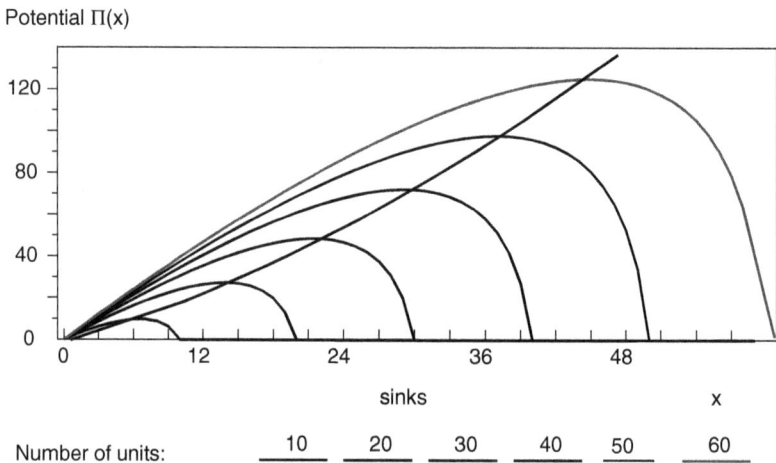

Figure V.5: A graphical representation of $\Pi(x)$, the potential of functional organization, x being the number of sinks. From left to right the curves correspond to 6 values of v, the total number of structural units $(10 \leq v \leq 60)$, i.e. the total number of sources and sinks, assuming that each structural unit is either a source or a sink. The graph joining the summits of these curves represents $F(v)$, the function *orgatropy*.

intuitively, we feel that these extreme situations are rather unsatisfactory, this may well be because of certain situations actually observed in human societies. In fact, there appears to be a close relationship between biology, the science of the organization of cells in living organisms, and sociology, the science of the behavior of individuals in a society. However, for the present, let us return to the idea of the potential of the functional organization of biological systems.

We may recall that the *state of organization* of a system was defined above by the distribution of its n sinks. For a given system, the state of organization represents a compromise between two tendencies indicated by the potential of functional organization, i.e. that of being composed, on one hand, of functionally independent elements, and on the other, of such highly specialized elements that they are totally interdependent. However, somewhere between these extremes, there must be a *particular* value of n, say n_M, which offers the best compromise, such that the potential of functional organization of the biological system is a maximum, i.e. *the "reservoir" of possibilities has attained its highest evolutionary development.* This would be in keeping with the great ideas of biological evolution, such as the accumulation of specific properties in living organisms, the survival of the fittest, and so on. The hypothesis we propose is of a fundamental nature since it formally expresses a certain functional equilibrium of biological systems, based on the accumulation of adaptive deviations, without which no living organism could survive. An example of this is the extraordinary adaptability of the human immune system through permanent confrontation with a hostile environment.

In any case, even if a suitable mathematical form could be suggested for the mechanisms underlying biological evolution, we would still have to determine which of the potential organizations of the system were effectively adopted during the evolutionary process. From what we have seen above, the organization actually observed is that which results from the increase in the stability of the hierarchical system during its development. Thus, there is surely a relationship between the potential of organization, corresponding to the O-FBS, and the hypothesis of self-association, which describes the gradual increase in the stability of the D-FBS. Is the functional organization, which increases the stability of the dynamic biological system, precisely that with the maximum potential of organization? Here we come upon a difficult problem indeed, and one for which we can offer no immediate solution. The difficulty arises because, for theoretical reasons, there can be no direct relationship between the potential of organization, which corresponds to a topological property of the system, and the property of stability, which depends on the dynamics of the system. These two properties of biological systems are complementary in the same way as the first and second principles of thermodynamics are in physical systems. Thus, the stability is a mathematical condition necessary for the life of a biological system, whereas the maximum potential of functional organization is an emanation of the evolutionary process. The study of the evolutionary component of living organisms — a novel concept closely related to the idea of the evolution of the species — may be expected to lead to new insights into the growth and development of physiological systems.

We may try to determine certain properties of the class of biological systems with a maximum potential of functional organization. The properties we seek, being comparable to those of physical systems for which the criteria of thermodynamic evolution are well known, should enable us to predict the evolution of these biological systems.

Biological systems tend to a maximum potential of functional organization

Let us consider the functional organization, i.e. the O-FBS or the topology of the biological system, having the greatest possible potential of organization and in which the distribution of the number of sinks for each functional interaction — or product — varies with time during its development. This corresponds to the *state of maximum functional organization*, noted n_M. Although this state may not actually be known, it is possible to determine certain properties of the dynamics of the

O-FBS[44] by analyzing the mathematical function representing the potential of organization of the biological system. Thus, as in the case of entropy discussed above, we may seek a *Lyapunov function* that would indicate the sense of variation of the state of maximum organization of the biological system. More precisely, it would indicate the zone of stability of Π, the function corresponding to the potential of organization, i.e. the region called the *attractor basin*, within which the value of the function must lie. Just as a marble released at the rim of a bowl will inevitably roll down to its equilibrium position at the bottom, the *position* of the state of the O-FBS in the phase space will tend to its attractor, i.e. the lowest point in phase space.

We may recall that the existence of a Lyapunov function is linked to the notion of the stability of a physical system. In particular, the function has a sign opposite to that of its derivative. The attractor basin consists of the phase space determined by the initial conditions of the dynamic physical system for which it will move towards a singular, locally stable point (Figure IV.2). Thus, the Lyapunov function allows us to visualize the time-variation of a system according to its initial conditions. Using an *extremum hypothesis* that defines a class of biological systems (Chauvet, 1993a), we have been able to show that *certain biological systems vary monotonously with time, such that the number of sinks either continuously decreases or continuously increases according to the initial conditions, until the systems reach the state for which the potential of organization is maximum* (Figure V.5).

We thus have the practical means of testing the hypothesis underlying the theory proposed. Experimental work should allow us to determine whether a given biological system satisfies the extremum hypothesis. The definition of functional interactions should make it possible to identify the sources and sinks in a biological system and determine whether the system possesses the property of the monotonous increase or

---

[44]The dynamics of the O-FBS, which refers to the dynamics of the sinks of the biological system, is distinct from the dynamics of the D-FBS, which refers to the dynamics of a physiological function or process.

decrease in the number of sinks, *even in the absence of any explicit knowledge of the dynamics of the O-FBS.* The theory proposed may then be subjected to further investigation in the usual scientific manner.

However, which of these alternatives — the increase or the decrease in the number of sinks — will actually be adopted by a biological system? This is a difficult question to answer since the mathematical solution of the problem depends on the initial conditions of the system, i.e. whether the number of sinks at the initial instant of development is smaller or greater than n_M. Furthermore, the initial conditions themselves depend on the stage of development of the biological system. However, there is much in favor of a system minimizing the number of sinks, i.e. maximizing the number of sources, since this would correspond to the maximum redundancy of the structural units. In fact, real biological systems certainly appear to possess the general property of being organized for *maximum reliability.* Thus the *optimum principle* would seem to be deeply rooted in the evolutionary mechanisms of life.

The developmental variation of a biological system is governed by its "orgatropy"

Let us now consider biological systems with a maximum potential of functional organization. The investigation of such systems should allow us to answer the fundamental question, posed at the beginning of this chapter, concerning the regulation of the functional organization of a biological system during its development. In particular, what happens when the modification of the number of cells in the organism leads to a change in the number of functional interactions?

This problem may be approached in terms of the variational aspect of the graph representing the topology of a biological system. We may recall that the principle of least action in a physical system is a variational principle (Chapter I). But how do we adapt this to the case of a biological system? What happens when, during the evolution or development of a biological system, the number of structural units, i.e. the degree of organization of a given physiological function, is modified? How does the system behave? How is it reorganized? Is there some force that obliges the system to follow one particular course rather than another? Considering the complexity of the problem, it would seem useful to break it down into simpler elements. Thus, let us first suppose that the structural units are reproduced with the sinks giving rise only to sinks, and the sources only to sources, such that the topological system remains unchanged. We shall subsequently see what happens when this condition is suppressed.

As we have seen, the potential of organization of a biological system depends on the number of sinks, which in turn depends on v, the total number of structural units at a given level of organization. We have introduced a novel mathematical function, $F(v)$, to describe the variation with time of the "reservoir" of the organizational possibilities of the system when the number of structural units undergoes a change. The value of $F(v)$ corresponds to maximum value of the potential of organization when this is considered as a function of the total number of structural units.[45] We have called this function "*orgatropy*", with reference to an organizational change in a *biological* system, on the model of the thermodynamic term "entropy", which refers to an organizational change in a *physical* system. Interestingly, the mathematical function representing the orgatropy of a biological system possesses the property of indicating the *direction* of the variation of the system with time. Furthermore, it can be shown that *if the degree of organization, i.e. the number of structural units, of a biological system at its maximum potential of organization changes — without any reorganization of the system — then the orgatropy of the system cannot decrease.*[46]

---

[45]$F(v)$, the orgatropy of the system is defined by: $F(v) = \Pi_{max} \circ h(v)$ where Π_{max} is the maximum value of Π, the potential of organization, and h is the implicit function defining the maximum value of Π, such that $d\Pi = 0$.

[46]This is mathematically expressed by the criterion of the variation of the system with time: $dF(v) \geqslant 0$, where dF is the differential of the function F.

In other words, the way in which the physiological products are exchanged between macromolecules, cells, tissues and organs of a biological system varies with time. Thus, the orgatropy, which represents the organizational "reservoir" of the system, *increases* when the number of macromolecules, cells, tissues or organs, undergoes a variation that is not accompanied by any reorganization of the system, the sinks subsisting as sinks, and the sources as sources. The criterion of the variation of the system with time, i.e. $dF(v) \geqslant 0$, determines the distribution of new structural units into the classes of sinks and sources such that the new value of the potential of organization remain as high as possible (Figure V.6). During the transition of the organization from the initial to the final state, the potential of functional organization *increases*, the variation

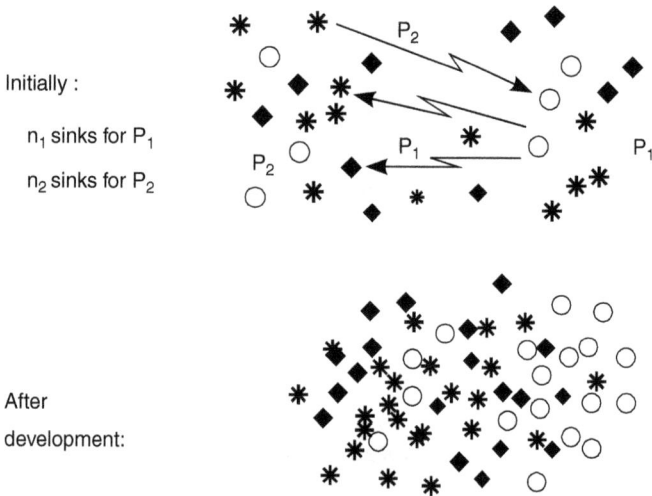

Initially :

n_1 sinks for P_1

n_2 sinks for P_2

After

development:

Figure V.6: The development of a biological system *without* the reorganization of its structural units. Initially, the units have common individual functions, e.g. n_1 units are sinks for product P_1, and n_2 units are sinks for product P_2. When the number of units increases by successive cell divisions or differentiations, the potential of functional organization maintains a maximum value while increasing. Thus, graphically, this value will lie on the orgatropy curve passing through the summits of the curves representing the function of potential organization for various values of the number of sinks (Figure V.5). In mathematical terms, we have: $dF \geqslant 0$.

with time being governed by the mathematical function of orgatropy.

The concept of orgatropy may appear highly abstract, but it is of a purely biological nature since it concerns the topology of biological systems. It implicitly includes the two fundamental properties of biological systems, i.e. that of the non-symmetry of sources and sinks and that of the maximum potential of organization. Because of the non-symmetry of sources and sinks in a biological system, the concept of *orgatropy* is completely different from that of *topological entropy*, which is a mathematical concept of an informational nature, and that of *physical entropy*, which is a thermodynamical concept of a structural nature. In effect, both topological entropy and physical entropy are mathematically expressed as: $S = -\rho\ln\rho$ where ρ represents a distribution of states of matter. In contrast, orgatropy plays the role of "functional entropy" for a particular level of biological organization. Furthermore, unlike the expression for topological entropy or physical entropy, the potential of functional organization of a biological system, i.e. $\Pi = \rho\ln(\upsilon - \rho)$, clearly possesses the property of non-symmetry. If the extremum hypothesis of the potential of functional organization is valid, then the O-FBS varies with time such that the orgatropy of the biological system increases. As we shall see below, the calculation of orgatropy is perfectly feasible when the functional couplings of the system are known, as in the case of the nervous system.

The reorganization of the structural units of a biological system leads to specialization

A fundamental property of living organisms, which we have not yet considered, is that of *functional specialization*, as observed for example in the organs of a body. During the development and growth of an organism, sets of cells associate in the body to form tissues with specific functions. Then, the various tissues are regrouped stepwise to form a structured organ with a well-defined physiological function. For example, the ventilatory function of the lungs requires the association of the lungs, the thoracic and abdominal muscles, as well as the system of respiratory regulation. The process that leads to the formation of the elastic walls of the lungs by the association of tissues made up of cells, all having the same structural properties, may be defined as that of *functional specialization.* In the framework of our theory of integrative biology, this process corresponds precisely to the creation of new organizational levels in a biological system, since each level is defined by a specific physiological function and is composed of structural units participating in this function. It would therefore be useful to formulate the property of functional specialization in mathematical terms.

Thus, the *functional specialization* of structural units may be defined as the development of these units for the *exclusive* emission of a given product. This process corresponds to the *partition* of the set of structural units. In mathematical terms, the partition of a set of elements corresponds to the decomposition of the elements into subsets possessing exclusive properties, i.e. an element belonging to a subset associated with a given property will not belong to any other subset with respect to the same property. In other words, if all the subsets are disjoined, each element of the set will belong to only one of the subsets. *The condition of functional specialization may then be expressed as the maintenance of this partition in time,* even when the topology of the system undergoes a modification because of a change in the distribution of sources and sinks (Figure V.7).

By introducing δ the *operator of hierarchization,* we have been able to show that, for a given class of systems with a maximum potential of functional organization, the mathematical condition of specialization leads to the emergence of a higher level of organization, such that: $\delta F \leq 0$, with $\delta v < 0$. The passage from the initial to the final states, which respectively contain $v^{(0)}$ and $v^{(f)}$ structural units, is accompanied, on one hand, by an increase in orgatropy with time, i.e. $\delta F(v) \geq 0$, because of the reorganization of structural units, and, on the other, by a decrease in orgatropy as a function of state, i.e. $\delta F \leq 0$, because of the emergence of a new level of

Initially:

State (0)

$n_1 + n_2 = n^{(0)}$

n_1 sinks for P_1

n_2 sinks for P_2

Partition:

Specialization:

State (f)

$n_1 + n_2 = v^{(f)}$

$n^{(0)} < v^{(f)}$

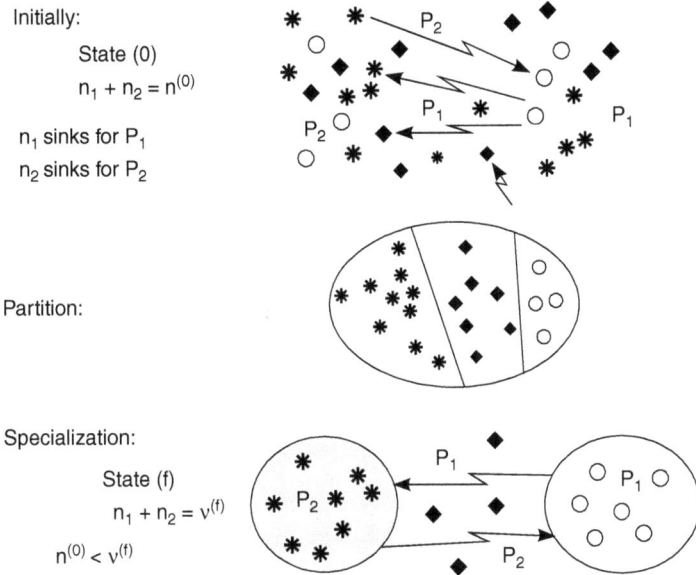

Figure V.7: The development of a biological system *with* the reorganization of its structural units. As in the case described in Figure V.6, the structural units initially have common individual functions, for example, n_1 units being sinks for product P_1, and n_2 units being sinks for product P_2. When the number of units increases by successive cell division and differentiation, the potential of functional organization maintains its maximum value while increasing. Graphically, the value of this potential will lie on the orgatropy curve, as shown in Figure V.5. Thus, here again we have: $dF \geq 0$, where d is the operator of mathematical differentiation. However, the new structural units are simultaneously specialized in the production of P_1 or P_2. In mathematical terms, the reorganization of the structural units of the system corresponds to a *partition* of the units, with the creation of new states, such that we have: $\delta F \leq 0$, where δ is the operator of hierarchization.

organization. This is analogous to the situation in thermodynamics in which the entropy of the physical system describes the variation of the system from the initial to the final states, without needing any knowledge of the actual path followed between the two states.[47] Thus, we see that orgatropy, as a

---

[47]This is possible because, mathematically, the differential of the entropy function is a total differential.

function of state, describes a variational property of a biological system.

Another new property is revealed by a biological system that satisfies the extremum hypothesis. Such a system reorganizes itself with the emergence of a higher level of organization specialized in the emission of a certain product. In other words, according to the definition of structural units, the system reorganizes itself by the creation of a class of structural equivalence. In short, *the double condition of the maintenance of the potential of functional organization at its maximum, on one hand, and the partition of the set of structural elements of the system, on the other, leads to a reduction in the number of structural units by the creation of classes of structural equivalence. Thus, a new level of organization is created in the hierarchical system.* Here, we have another variational property of the orgatropy of a biological system.

With these two variational properties of orgatropy in mind, we may now formulate our problem as follows: Is there an extremum principle that would allow us to combine these variational properties? In other words, can we define a new state function that will indicate the direction of the evolution of a biological system, taking into account the *specialization* as well as the *hierarchization* of its structural units?

Optimum development implies increased functional order and decreased orgatropy

According to the extremum hypothesis, the variation of a biological system during its development is such that the functional couplings between sources and sinks correspond to the maximum value of the potential of organization. It is rather remarkable that the mathematical properties of this particular class of formal biological systems should fit in so well with biological reality. In effect, if no constraints are applied to such systems, then the orgatropy increases; and if the constraint of the specialization of structural units is imposed, then the system undergoes reorganization with the emergence of a higher level of organization accompanied by a decrease in orgatropy.

These are precisely the two properties we need to describe the general behavior of a biological system.

Nevertheless, it is difficult to admit that a complex biological system might be governed simply by means of its orgatropy, which, in spite of successive reorganization of the system, is bound to increase and eventually lead to instability. We therefore need a coherent description of a highly organized functional organization that will take into account all the available knowledge concerning living organisms and, in particular, the two properties mentioned above. This new state function will have to include, on one hand, the variation of the entropy of the system during its development, and, on the other, the effect of the specialization of its structural units. In other words, *the state function will have to ensure that the system maintains a maximum potential of functional organization while reorganizing itself.* Moreover, it will have to describe the variation of the topology of the biological system as it undergoes changes in the number of structural units and modifications of the functional couplings between them. For example, if we consider the physiological function of learning, the number of synapses in the nervous system decreases during the development of an organism, while large numbers of inter-neuronal associations are established. These are such highly complex phenomena that we may wonder whether a single mathematical function could possibly explain them all.

Here again, an appropriately formulated Lyapunov function offers a promising top-down approach. We have therefore constructed a mathematical represention, called the *"functional order"* of the biological organism, which consists simply of the sum of the orgatropies at each level of organization of the system. Then, with the condition of the specialization of structural units, the organization of the biological system varies with time, satisfying the mathematical relationships governing the degree of functional organization and the number of sinks involved (Chauvet, 1993a). Briefly, these relationships may be stated as follows:

(1) For each elementary physiological function, no more than two new units are created, each being specialized in the

elaboration of a particular product, thus leading to the emergence of a higher level of organization;

(2) The *functional order* of the system increases, while its *orgatropy* decreases; furthermore, the potential of functional organization of the system decreases though it remains a maximum; and

(3) When the system attains its greatest degree of specialization, the highest level of organization contains no more than two structural units involved in the elaboration of a given product.

The organization of physiological functions thus satisfies an optimum principle, based on the potential of functional organization, and leads to a theoretical law governing the *functional order* of the system. *Thus, in addition to its physical structure, which varies with time according to the thermodynamic principle of entropy, the living organism can be considered to have a functional component that varies with time according to the biological principle of orgatropy.* If this theory were substantiated, it would make a significant contribution to the problem of distinguishing the living from the non-living.

The potential of organization of the nervous system is maximum during its development

The functional interaction between two neurons is, in fact, an interaction between a neuron that acts as a source and several synapses, situated on target neurons, which act as sinks. Therefore, the calculation of the potential of functional organization, i.e. $\Pi = \Sigma[sinks]\ln[sources] = \Sigma n \ln(\upsilon - n)$, is somewhat more complicated than in the case of a simple source-to-sink interaction. The "product" of the interaction between neurons corresponds to the electric potential transmitted between the neurons.

We may test the theory of orgatropy, outlined above, by applying it to the growth of nervous tissue when the number of synapses, i.e. sinks, varies in a system containing a given number of neurons (Figure V.8). According to the theory, the value of the potential of functional organization must then be a maximum. But does the system belong to the class of systems satisfying the extremum hypothesis? In other words, do we observe a monotonous variation of the number of sinks?

Let us first consider the theoretical aspects of the two distinct functional processes involved in the development of nervous tissue. The first process concerns cell proliferation during which there is an increase in the number of neurons accompanied by the increase in the orgatropy of the system from its initial value to a maximum, without any reorganization of its structural units. The second process concerns the reorganization of synapses. During this process, there is a decrease in the number of target neurons accompanied by a reorganization of synapses such that there is a decrease in the number of sinks accompanied by a decrease in the orgatropy of the system.

Now, are these theoretical findings in keeping with results of observations? It has been determined that, during the interval of time in which the neurons proliferate to reach their maximum number and their extensions attain their targets, cell death takes its toll by cutting back the number of neurons to about half the maximum value. Another phenomenon that intervenes is that of specialization (Figure V.8), illustrated by the elimination of supernumerary innervations, as observed, for instance, in the submandibular ganglion of the rat (Brown, Hopkins and Keynes, 1991). These phenomena correspond to the two functional processes described by the theory: on one hand, the neuronal and synaptic growth necessary for the increase in the number of synapses and, on the other, the monotonous decrease in the number of synapses corresponding to the reorganization required, for instance, by the mechanism of learning. Thus, the potential of functional organization appears to maintain a maximum value during the growth and development of a living organism.

What distinguishes the living from the non-living?

One of the great riddles of life may be stated as follows. Given the position and the momentum of each molecule of a living

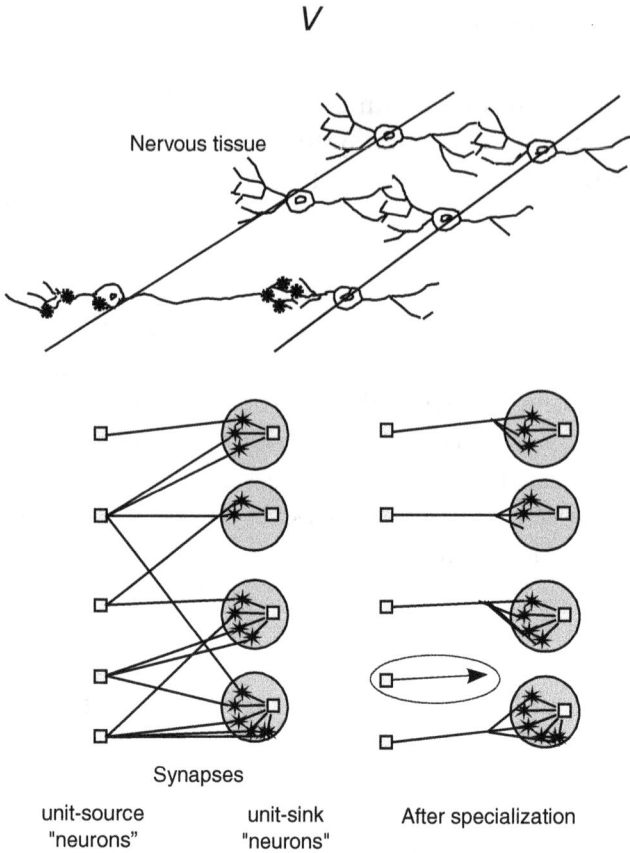

Figure V.8: A schematic representation of the development of the nervous system. The top of the figure shows nervous tissue composed of several layers of interconnected neurons. The bottom of the figure shows two of these layers with functional interactions *via* the set of synapses, before specialization (on the left), and after specialization (on the right). The theoretical model fits biological observations, since the increase in the number of neurons (without any reorganization) is associated with a decrease in the number of synapses (specialization with reorganization). These phenomena correspond to two distinct functional processes.

creature at a certain instant, would we be able to reconstitute the biological organism? Of course, with adequate knowledge we might be able to reproduce the *physical* structures of the organism. However, these structures would further need to be organized to produce *functional interactions*. As a simple physical structure, a molecule with its movements of rotation and translation, under the influence of various forces, may interact with other molecules, but it will have no real *function*. To execute a function, in the *biological* sense, it would have to produce a non-local transformation, i.e. it would have to move across a structural discontinuity to accomplish a specific task.

We have, in the preceding chapters, described the living organism in terms of its functional interactions, and explained how the biological system develops according to an optimum principle. We have also seen how the functional organization of a biological system varies with time following certain structural modifications. Thus, as the living organism develops, the orgatropy of the biological system, considered as a state function, decreases while its functional order increases. Moreover, the extremum condition concerning the potential of functional organization, and its corollary, i.e. the monotonous variation of the number of sinks, leads to an increase in complexity at a given level. This complexity can only be decreased by a reorganization of the biological system such that the number of hierarchical levels increases. Finally, compared to non-living matter, in which the thermodynamic entropy of a physical system cannot but increase, thus leading to an increase of molecular disorder, the living organism is specifically characterized by the execution of physiological functions that decrease the orgatropy of the biological system, thus leading to an increase in its functional order. All these findings, summed up Table V.1, show that while living organisms *contain* the essential mathematical properties of non-living matter, they also possess *additional* mathematical properties corresponding to the phenomena of life.

The theory of integrative biology we propose is based on an original approach to the structural and functional organization of living organisms. On one hand, the *structural organization* of a biological system may be explained by the fundamental physical concept of *force*. As we shall see in Chapter VI, the force acting between two structures can be interpreted by quantum field theory in terms of an exchange of *virtual* particles between the structures. This interpretation incorporates the principle of action and reaction, characterized by the property of *symmetry*. The collisions between the particles can be

Table V.1: Mathematical criteria of the distinction between living organisms and non-living matter.

Organization	Theory	Mathematical functions	Characteristics	Significance
Structural organization				
Non-living matter *and*		Thermodynamic entropy	Increasing	*Increase in structural disorder*
Living organisms	Geometry	Neguentropy	Decreasing	
		Lyapunov function	Positive or null	
Functional organization				
Living organisms	Geometry and Topology	Orgatropy	Negative or null	Creation of hierarchical levels
		Functional order	Positive or null	*Increase in functional order*
		Lyapunov function	Positive or null	

described by the *entropy* function, from which we may deduce a Lyapunov function to determine the equilibrium state of the geometry of the system. On the other hand, the *functional organization* of a biological system, based on functional inter-actions corresponding to the unidirectional transfer of sub-stances or signals between the structural units of the system, is characterized by the property of *non-symmetry*. The functional interactions, which are, in a certain sense, analogous to molec-ular collisions, can be described by the *orgatropy* function, from which we may deduce a Lyapunov function to determine the equilibrium state of the geometry as well as the topology of the system.

We may therefore imagine that, in a living organism, there are symmetrical collisions between the particles of the physi-cal system and non-symmetrical interactions between the structural units of the biological system. This is precisely the point of divergence between the physical theory usually applied in biology and our theory of functional biological organization. Finally, a *living organism is characterized by two complementary sets of mathematical state functions: one gov-erning its physical structure according to an optimum princi-ple of thermodynamic entropy; and the other, which may be expressed as a law of functional order, governing the organi-zation of its physiological functions.*

Mathematical abstraction can lead to new insights in biology

The theory of functional physiological organization proposed here may appear rather forbidding at first sight, perhaps because of the unusual degree of mathematical abstraction involved. To be sure, it will need to be greatly refined for prac-tical applications in biology and many additional concepts may have to be developed. One of the advantages of the theory is that it is based on a unique, extremum hypothesis, which states that *the organization of a functional biological system varies with time such that its potential remains at a maximum.* Now, a theory may be refuted by one of two methods, i.e. either by proving the initial hypothesis invalid, or by demonstrating the failure of its predictions. Thus, with the first method, we could test the observable corollary of the extremum hypothesis, according to which the distribution of sinks in a biological system varies monotonously with time during its development. With the second method, we could examine the consequences of the extremum hypothesis on the organization of physiological functions. We could also deduce specific properties from the mathematical function corresponding to the potential of functional organization, e.g. the decrease in the complexity of a biological system during successive functional reorganizations, accompanied by an increase in specialization at each higher level of its hierarchical organization.

However, the major advantage of our approach is that it allows us to consider a living organism in terms of a biological system, thus giving us *an integrated image of the underlying reality*. Thus, we have been able to retrieve two essential properties of living organisms directly from the extremum hypothesis. The first property is related to our intuitive perception of biological evolution. According to our theory, the biologist's idea that evolution tends to produce ever more complex forms of life corresponds to the constant search of a biological system for optimum organization in terms of the potential of functional organization, which may be interpreted as *functional complexity*. Thus, any perturbation, due to the environment or to an internal mutation, will lead to the reorganization of the biological system such that its functional order increases.

The second property is related to our perception of biological order and organization. The emergence of higher levels of functional organization corresponds to greater specialization of the structural units, i.e. to the hierarchical organization of the system. From an intuitive point of view, a biological organism appears to be better "ordered" the greater the number of specialized organs and levels of organization. However, as the level of organization rises, the number of functional interactions between the structural units decreases, reducing the degree of complexity. This might explain why, at the highest level of biological organization, there are generally only two organs for each physiological function. It is certainly remarkable that, after all the innumerable *cell* divisions and differentiation during its growth and development, the organism ends up with *pairs of organs*, such as the eyes, ears, lungs, kidneys, and so on. Since the maximum potential of functional organization is closely linked to the reliability of the system, a minimum redundancy of two units may be necessary at the highest level of biological organization. There are of course some obvious exceptions, for example, the digestive tube, but from the mathematical point of view, the internal space being in continuity with the external space, this structure may be assimilated to that of a tore. In other words, the human body, for instance, is topologically equivalent to a tire, or a donut.

Another advantage of our approach is that it provides a criterion of biological evolution that is compatible with the emergence of new, higher levels of organization accompanied by greater specialization of the structural units of the organism. It also explains the development of a biological system by means of successive, bi-unitary specialization and association, accompanied by a reduction of the potential of functional organization, ensuring the conservation of its physiological functions. This leads to the principle of vital coherence in its restricted form. Among all the organizations offered by the "reservoir" of developmental possibilities that could favor the establishment of certain couplings, it is precisely the organization observed that ensures an increase in the functional order of the biological system. We may therefore ask *why* a particular state of organization, i.e. that with the maximum potential of organization, should be observed rather than some other, which might be equally favorable. This intriguing question is of the same nature as that of the origin of the physical attraction existing between two material bodies. Does the attraction exist because it is necessary that the stability of the physical system increase? While the quantity representing the potential of organization of a biological system does possess properties allowing a convenient formal representation of living organisms, there may well be other possible approaches. The method we have proposed has the quality of great simplicity and this, as we have seen, is an important characteristic for a sound theory.

Of course, the abstract vision of living organisms should not be allowed to lead us to neglect biological phenomenology. We should constantly ask to what extent the results obtained with a formal biological system might be extended to a real biological system. Our integrated theory of functional organization has two complementary components: one is the conceptual framework representing a living organism in terms of non-symmetrical functional interactions between its structural elements; the other is the criterion of biological evolution based on a hypothesis, involving the maximum potential of functional organization of a living organism, that requires to be experimentally tested. It is certain that the theory as it stands will be considerably enriched by the incorporation of specific results obtained from real biological systems. Finally, it would seem reasonable to hope that a plausible explanation of the integrated functional organization of living organisms may emerge from the discovery of optimum principles in biology.

VI

THE ROLE OF SPACE IN FUNCTIONAL BIOLOGICAL ORGANIZATION

A Biological Field Theory

VI

THE ROLE OF SPACE IN FUNCTIONAL BIOLOGICAL ORGANIZATION

A Biological Field Theory

The physiological activity of a living organism depends directly on the functional interactions between the structural units of the body. For instance, a signal emitted by a neuron is propagated towards other neurons, and a molecule synthesized by an endocrine cell is carried by the blood circulation to target cells situated at a considerable distance from the site of production. Again, during the growth and development of an organism, signals emanating from cells in a given tissue provide information to neighboring cells concerning their relative positions. And at a higher level of organization, for example in the respiratory system, the muscular activity of the diaphragm produces the difference in pressure causing air to flow in the bronchi and the alveoli of the lungs.

Thus, the functional interactions in a living organism may be of considerable variety and complexity. However, whether it is a signal that is propagated, or a molecule that is transported, or even the influence of some factor that may not be directly observable, all functional interactions between the structural units of the body correspond to the transmission, at finite velocity, of an activating or inhibiting signal from a source to a sink. The finite velocity of transmission naturally refers to the *space* the signal travels through in a certain time. As we have seen, this space corresponds to the hierarchical structural organization of the body. In Chapter V, we discussed the dynamics of the functional organization, the O-FBS, during the growth and development of a biological system. Now, how does the O-FBS affect the dynamics of the physiological functions in a given functional organization in the adult state, i.e. when its development is completed? In this connection, let us first recall some fundamental physical concepts and distinguish them clearly from the novel biological concepts we propose to introduce.

Symmetry and locality are essential aspects of physical fields

In physics, the field concept is generally used to explain the interactions between bodies or particles. For example, a *scalar* field represents the potential energy of a body, which depends on its position. This is the simplest case, since it involves the consideration of magnitudes only. *Vector* fields, such as electric, magnetic and gravitational fields, represent physical quantities associated with magnitude as well as direction. Such fields involve a high degree of mathematical abstraction but the phenomena they describe are familiar to most people who have ever watched TV, used a cellular telephone or ridden a roller coaster. However, the field concept is much more than just a convenient way of representing physical phenomena. It has led to the construction of a physical *field theory* that aims to describe the properties of all physical fields, i.e. the electric, magnetic, gravitational and nuclear fields, within the framework of a unified system. Although this unified field theory remains to be fully verified, it has already proved extremely productive. Let us recall briefly some of the steps that have led to this great new idea in physics.

According to one aspect of Newton's third law of motion (1687), the force of attraction between two bodies is inversely proportional to the square of the distance between them, whatever the position of the bodies in space. As soon as the law was published, the mystery of the influence of attraction, acting invisibly across space, aroused much controversy. However, at the end of the seventeenth century, Lagrange (1736–1813) introduced one of the most potent concepts of physics. The essence of the idea was that if the effect of an action appears at all points of space, this is due to a gravitational potential, i.e. to a gravitational field existing at each point of space. Lagrange thus replaced the concept of an action producing an effect at a distance by that of a potential. Finally, the field concept, abstract as it was to begin with, has finally turned out to be of a real, quantitative nature.

What does all this actually lead to? According to Newton, the time-variation of the force of gravity is continuous, from one instant to the next, but the space-variation occurs between two points that may be very far apart. According to Lagrange, the gravitational potential varies continuously in time and in space, at all points of space. And according to the expression of the law of gravitation based on the principle of least action (Chapter I), a body moves from one point of a curve to another such that a certain quantity, called the action, which is equal to the difference between the kinetic and potential energies of the body, is minimum. This means that among all the possible trajectories, that chosen by the body will minimize the action. These three formulations of Newton's third law of motion are mathematically equivalent, yet in each case the physical significance is quite different. However, does the law express *action at a distance*, the influence of a *local field,* or an *extremum principle?* Here again we come upon the strange character of the scientific interpretation of physical reality and the intelligibility of nature. Indeed, none of the forms of Newton's third law of motion can tell us *why* the attraction between two bodies exists; it can only tell us *how* the attraction works.

Finally, Einstein's theory of general relativity explained the attraction between bodies in space by linking the concepts of geometry and gravitation. This theory, truly a monument of human thinking, is highly abstract and can only be understood by following a mathematical reasoning based on tensor

VI

calculus. The bases of Einstein's special theory of relativity (1905) are identical to those of his general theory (1915). However, the former allows the formulation of physical laws that are valid in any inertial frame of reference, i.e. for all observers in unaccelerated motion, whereas the latter requires that the laws be valid in a non-inertial coordinate system, i.e. a system in which one referential may be accelerated with respect to another. As Einstein put it:

> "The simpler and more fundamental are our suppositions, the more complicated becomes our mathematical reasoning. The path from theory to observation becomes longer and more delicate. As paradoxical as it may seem, we may say that modern physics is simpler than it used to be and therefore appears more complicated. The simpler our image of the external world and the more closely it incorporates facts, the better it reflects the harmony of the universe."

Einstein's theory of relativity literally transformed our understanding of the physical universe by giving us an entirely new vision of Newton's theory of gravitation. In fact, the general theory of relativity actually leads to a new interpretation of the world we live in. We abandon the mechanical concept of Newton's law that expresses in "simple" terms the relationship between the motion of a given body and that of another body, situated elsewhere at the same instant, and therefore in the same coordinate system. Instead of this, we adopt a new law governing the motion of these bodies in different coordinate systems, which may also be moving relatively to each other. The law of motion that is valid in all systems of space and time coordinates is expressed by the equations of general relativity. However, these equations are not involved in inertial coordinate systems. The geometry of space defines the nature of the coordinate system, and the movement of the body (i.e. its mass and velocity) determines the system of coordinates. Therefore *it is the geometry of space that produces the force of gravity*. To be sure, this is a rather simple way of summing up the fundamentals of the general theory of relativity. In mathematical language, we may say that since the coordinate system defined by the space-time interval is non-inertial in curved space, the geometry of the space is non-Euclidean. The powerful idea of formulating the laws of nature to take into account all possible coordinate systems, inertial and non-inertial, led to a new geometry according to which the motion of masses produces a curvature in space. How does this fascinating geometrical interpretation of the universe affect the concept of physical fields? Before trying to answer this question, it may be useful to describe the *electromagnetic field*.

Maxwell's equations (1872) brilliantly brought together the phenomena of electricity, magnetism and light in a single set of mathematical relationships that laid the foundation for electromagnetic field theory. The concept of the electromagnetic field originated in the pioneering experimental work of Oersted (1777–1851) and Faraday (1791–1867).

In Oersted's experiment, an electric current circulating in a loop of copper wire was shown to deviate a magnetic needle. In physical terms, the electric current may be said to exert a force on a magnetic pole. However, the quantification of this force, i.e. the mathematical expression of the force in terms of the parameters of the electric circuit, the relative positions of the copper loop and the magnetic needle, and so on, is not quite as simple as in the case of Newton's law of falling bodies. Nevertheless, it is somewhat easier to analyze the phenomenon when the needle is moved in a plane perpendicular to that of the copper loop. The needle may be assimilated to a segment of a straight line, and when the needle is moved in a plane, its center describes a curve. Thus, at each point of the curve, the straight-line segment is tangential to the curve. What does this curve represent? It is, in fact, an abstract, imaginary line, which acquires a certain reality only when a "signal receiver", i.e. the magnetic needle, is placed on one of its points. Each point on the curve is associated with a definable property, called the "function of the point", given by a set of numbers representing the direction and the length of the force vector. The space in which the needle is moved may then be thought of as a set of points identified by the numbers that characterize the property under investigation. This is what constitutes a *field*. Thus, when the force studied is that of gravity we have a gravitational field, and in the case of magnetism, we have a magnetic field, and so on. The magnetic field is a set of lines of force that can be physically materialized in a plane, for example, by means of iron filings on a sheet of paper.

Figure VI.1a illustrates Oersted's experiment in which an electric current produces a magnetic field in a coil, or a solenoid, made up of several circular loops. If the solenoid were replaced by a bar magnet, the lines of force observed would be similar. In this case, of course, the source of the field would be the bar magnet. Thus, we may say that the properties of the solenoid are identical to those of the bar magnet if these properties are represented by the magnetic field. In other words, the knowledge of the properties of the field is all that is required for the description of the phenomena observed, and the differences between the nature and the quality of the sources are of no importance. This is true in the case of a physical field but, as we shall see below, the situation is fundamentally different in the case of a biological field, in which the properties of the interacting sources and sinks play an essential part.

The main result of the analysis of the experimental phenomena described above is that a magnetic field is always associated with a flow of electric charges. But, as we have seen, a static electric charge produces an electric field by itself. This is common experience for anyone who, in cold dry weather, has ever used a hair-comb and noticed the crackling sparks produced, or stepped out of a car and received an unpleasant electric discharge from the door handle. Furthermore, an electrically charged sphere produces an effect closely analogous to that of a gravitational field. If the sphere is immobile, Coulomb's law mathematically expresses the force acting between the sphere and another charged particle in its neighborhood. Now, Coulomb's law and Newton's law of gravitational attraction are very similar, and this of course is explained by the similarity between the electric and gravitational fields. The field produced by an immobile electric charge is called the electrostatic field. Thus, an electric charge placed in the electrostatic field of another electric charge will be subjected to a force according to Coulomb's law.

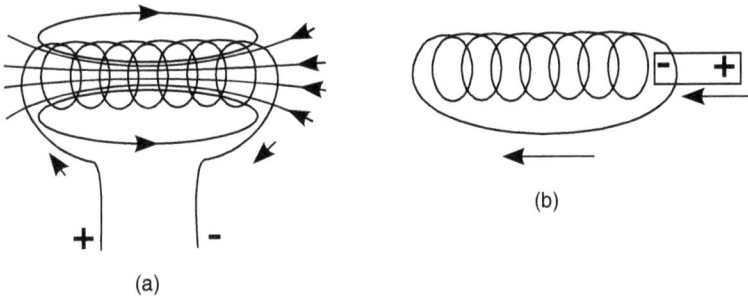

Figure VI.1: Electromagnetic field theory emerged from experiments carried out by (a) Oersted, and (b) Faraday.

A static electric charge produces an electric field by itself, and a moving electric charge produces a magnetic field; in other words, a variation of the electric field produces a magnetic field. However, is the converse true, i.e. does the variation of a magnetic field produce an electric field? Faraday (1791–1867) experimentally demonstrated the existence of induced electric currents. In his experiment, the movement of a magnet leads to an induced electric current in a coil (Figure VI.1b). In Oersted's experiment, the moving electric charge produces the magnetic field. In contrast, in Faraday's experiment, the moving magnetic charge produces the electric field. Thus, these two experiments beautifully illustrate the symmetry of the electric and magnetic fields. However, if we consider an electric current flowing in a coil, it will produce a magnetic field, which in turn will induce an electric field, which again will produce a magnetic field, and so on. It is therefore far more convenient to consider electric and magnetic phenomena in the framework of a unified electromagnetic field.

The experiments described above demonstrate that an electromagnetic field produces a force that can lead to movement. Since this implies the expenditure of energy, we may consider the electromagnetic field as a reservoir of energy. Thus, the field is not merely an abstract concept that is useful for describing observed phenomena but, in fact, it possesses a *physical reality*. The behavior of the electromagnetic field is fully described by Maxwell's equations, expressed as a set of four partial differential equations. The importance of these equations lies in the fact that they provide not only an interpretation of all the experimental work in electricity and magnetism but also a representation of the structure of the electromagnetic field. Maxwell's equations represent the dynamics of the interactions between the electric and magnetic fields *independently* of their origin. In other words, the equations attest to the reality of the electromagnetic field. This concept of physical nature has profound implications. Thus, space is literally suffused with electromagnetic energy since, once created, the electromagnetic field is propagated indefinitely and distributed in space. The field may be considered to act locally on a sink in this space, in the sense that the field at a neighboring point exerts an influence on the sink. This phenomenon is therefore governed by *local equations* since we can use the knowledge of the field at a given point to determine its action at the neighboring points.

The solutions of Maxwell's equations are of extraordinary significance. One particular solution corresponds to the *electromagnetic wave*. From the theorist's point of view, an electromagnetic wave corresponds to a *plane wave* that satisfies all four equations in response to an oscillating charge. In simpler terms, it represents the variations of an electromagnetic field that is propagated in space, the variations being a succession of changes in the electric field producing a magnetic field, one field producing another a little further in space and a little later in time, the process being repeated indefinitely. Where are the lines of force situated? The theory shows that the lines of force lie in a moving plane that is perpendicular to the direction of propagation. This is why the propagated wave is called a plane wave. If the charge ceases to oscillate, then the field becomes an electrostatic field. And here we have another important solution of Maxwell's equations.

From the experimentalist's point of view, electromagnetic waves have a very real existence. Fifty years after the theory was established, Hertz (1857–1894) demonstrated the existence of electromagnetic waves traveling at the speed of light.

Astronomers and astrophysicists have since confirmed that space is permeated with electromagnetic waves, and that some of the waves we receive today originated in the explosion of stars situated at the outermost reaches of the universe. Could we imagine any more spectacular substantiation of Maxwell's ideas than the fact that we can actually measure the energy carried by waves that have traveled through space for billions of years, even long after the source that produced them may have disappeared? The concept of the field may be rightly considered as one of the more important concepts of modern physics since it allows us to describe the interactions between entities, even when these may have ceased to exist.

Having sketched out some of the main aspects of field theory in physics, let us now see what distinguishes biological fields from physical fields.

Non-symmetry and non-locality are essential aspects of biological fields

Summing up the fundamental idea of biological organization presented in the preceding chapters, we may say that the functional organization of a biological system can be represented by a mathematical graph in which the summits correspond to the sources and sinks of the system and the arcs correspond to the functional interactions between them. The functional organization is the general framework within which all the physiological processes of the living organism are carried out. These physiological processes are under at least four types of constraint: (a) the *structural and functional hierarchies*; (b) the *geometry* of the system that leads to time lags in the transmission of signals; (c) the *structural discontinuities* that physically compartmentalize the physiological processes; and (d) the *spatiotemporal continuum* within which these processes operate.

This clearly poses a new problem in biology: How can the constraints mentioned above be introduced into the representation of dynamic biological processes as they vary with time? To solve the problem, we have adapted the physical theory of fields to the specificities of biological phenomena. This has led to the construction of a novel formalism, called the formalism of *structure propagators* (*S-propagators*, in short), which allows the representation of physiological phenomena in a space-time continuum while taking into account the distribution and the hierarchy of the structural units in space, as well as the structural discontinuities of the organism (Chauvet, 1990).

As we have seen, in contrast with physical phenomena, which are generally of a *local* nature, biological phenomena are by their very nature *non-local*. The property of non-locality, together with that of non-symmetry, contributes to the specific character of biological systems. *In a biological system, the role played by non-locality in its dynamic organization is quite as essential as that played by non-symmetry in its functional organization.*

The transport of the product of a functional interaction by biological fields corresponds in fact to the transport of the product of a non-local interaction associated with a local transformation. In effect, if the product of a functional interaction such as a hormone, destined to modify, for example, the molecular concentration of some substance or some other physiological parameter, moves from a source to a sink, then there must exist some mechanism of transport, e.g. a pressure gradient or an ionic gradient. Thus, the concept of the functional interaction involves, in mathematical terms, the action of a *field operator* on a variable emitted by the source. More precisely, the field operator acts on a variable, the value of which undergoes a transformation because of the source.

The field operator is conveniently represented as shown in Figure VI.2. Since the source and the sink are in different places in physical space, the field operator actually transports the quantity ψ over a certain space during a certain time. The relationship between the field operator $\mathbf{H}$, the field variable ψ, and the source Γ may then be written in the form of a simple equation: $\mathbf{H}\psi = \Gamma$. This equation describes the local variation with time of the quantity ψ, which may be, for instance, the concentration of a certain molecule in the space in which it is transported. This type of formalism has already been used to describe phenomena in the nervous system. Beurle (1956) was

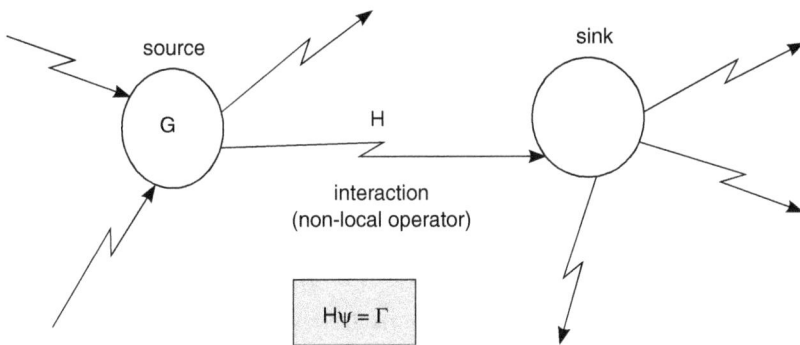

Figure VI.2: Diagram representing a biological field with a non-local, non-symmetric operator **H**. The field moves from one point to another under the action of this operator. The mathematical properties of the operator are the same as those of the functional interaction.

the first to investigate the properties of a group of neurons by using a field representation of nervous activity, and Griffith (1963) put these ideas in a more elaborate mathematical form. However, these authors considered only one level of biological organization, i.e. the neuronal level. The biological field theory (Chauvet, 1993e), taking into account several levels of organization, offers all the advantages of a field theory in physics. As in the case of electromagnetism and relativity in physics, the reciprocal effect between the source and the biological field leads to the propagation of the field in biological systems. The mathematical form of the theory is quite simple. We may recall that an equation containing variables of space and time is a partial differential equation. This representation of physiological phenomena in space and in time, based on the specific biological concepts of non-symmetry and non-locality, constitutes the essential element of our biological field theory (Chauvet, 1993c).

Thus, from the mathematical point of view, the functional interaction involves a source, the variable "transported", and a sink. The corresponding biological phenomenon consists in the emission by a cell of a product, which may be a molecule or a signal, its transport over a distance and its action on a target cell, followed by a transformation of the product, leading in turn to a process of emission. Viewed intuitively, this sequence of events consists of a repetition of elementary processes operating between sources and sinks, in other words, between one point and another, *within which transformations occur.* From a more scientific standpoint, we might say that these processes correspond to non-local transformations, as opposed to local transformations. This description brings out a fundamental characteristic of living organisms, i.e. *the existence of non-local interactions associated with local transformations.* The essentially geometrical nature of this characteristic is incorporated even in the terminology used. But how could these transformations possibly occur *within* a point, which is by definition dimensionless? The question is now easily answered since, according to our theory, the operator involved in the transport of the product must be non-local and, for this reason, it must include a *propagator,* in other words, a space function that takes into account the specific nature of the transport, which may be active or passive, enhanced or attenuated, and so on. The diagram in Figure VI.2 represents a biological field under the influence of a propagator (Chauvet, 1999).

The property of non-locality is inherent to biological systems

In physics, the property of non-locality, which arises from the fundamental ideas of quantum theory, has the most extraordinary consequences. Thus, the photon, a quantum of electromagnetic radiation, assimilated to a particle with zero rest mass and zero charge, has the surprising ability to pass simultaneously through two slits, i.e. it appears to be in two places at once. However, this example of physical non-locality, which is not represented in the mathematical equations of quantum theory, is an exception. In contrast, the property of non-locality in biological systems appears to be very general. This novel concept has been deduced entirely from theoretical considerations, such as those of the hierarchy and the continuity of biological systems. Figure VI.3 shows how the property of non-locality may be physically represented in a very general

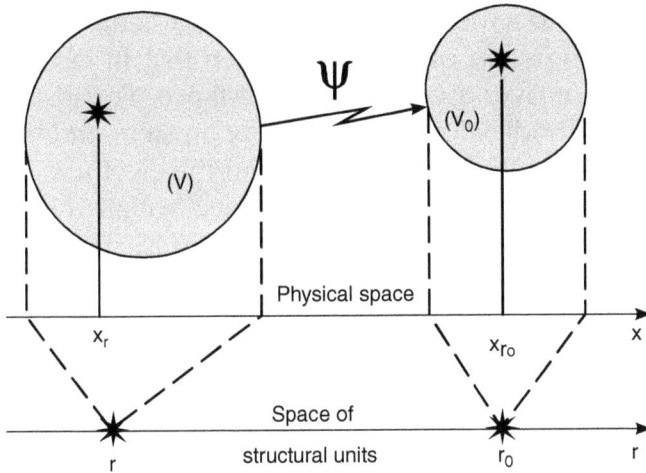

Figure VI.3: The property of biological non-locality reflects the interaction between two *volumes* in physical space (above), rather than two *points* in the space of structural units (below).

way. In this representation, a structural unit at a point **r** in the space of structural units is in fact a volume (*V*) in ordinary physical space. This illustrates how difficult it is to represent the elementary biological function, i.e. the action ψ operating from one point on another, which in fact corresponds to a phenomenon occurring in ordinary physical space, in terms of *the action of one volume on another.*

In a continuous representation, the structural units are localized in space according to their density. Thus, at each point of Cartesian space, the density indicates the number of structural units at a given point or, more precisely, about this point in an infinitesimal interval of the space. In a hierarchical system, we would have to consider the density of structural units in the space of structural units (as represented in Figure II.6), rather than in physical space. Since in this hierarchical system, a point of space corresponding to groups of units is in itself a space of units, the distance between two points may be infinitely small at a given level, but finite at a lower level. An example taken from the physiology of the nervous system may make this idea clearer. Thus, the propagation of a field from one neuron to another, both of which may be considered to be very close together, in the sense of mathematical continuity, is the expression of the propagation that in fact occurs in physical space over a considerable distance. For instance, the neuronal cell bodies localized in the brain stem and the spinal cord are connected by long axons, and the electric potential is propagated over a considerable distance within the neurons. However, in terms of functional interactions, the field *in the specific space of the neuronal network* is propagated from one neuronal cell body to another that is infinitesimally close. Let us now investigate this type of propagation in a hierarchical biological system.

How are biological processes propagated in space?

Experiments on the biochemical mechanisms of living organisms are generally based on the assumption that the media investigated are perfectly homogeneous. Of course, this is far from being the case in reality since biological systems cannot be reduced merely to phenomena occurring in a test tube. This is why we propose to introduce the concepts of time and space in biology. From this point of view, the living organism is considered as a set of media, corresponding to the spaces of structural units, within which specific physiological processes are carried out. The concept of the *structural discontinuity* allows us to describe physiological phenomena in which the functional interaction is propagated across different media. Thus, between two biological structures situated at a distance, there exists another structure such that there is a coupling between the dynamics of different media. A simple example of this is given by molecular transport of a molecule from one medium to another across a membrane. The presence of a structural discontinuity between the source and the sink means that the interaction ψ will have to occur at a lower hierarchical level in the form of another interaction ϕ, as illustrated in Figure VI.4.

Figure VI.4 represents the propagation of an interaction ψ between two points, r_i and r_0, situated at, say, level 2 of the organization. To move from r_i to r_0, for instance from one cell

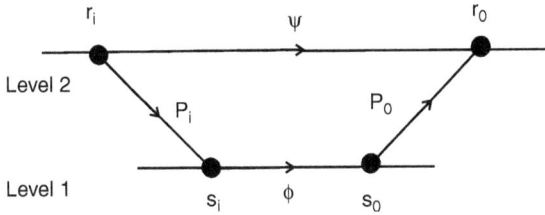

Figure VI.4: A functional interaction ψ operating between a source situated at r_1 and a sink situated at r_0 must first shift from level 2 to level 1, where it acts in the form of interaction ϕ, before returning to level 2. It crosses the structural discontinuities at r_1 and r_0 under the action of the operators $\boldsymbol{P}_1$ and $\boldsymbol{P}_0$.

to another, the interaction crosses a discontinuity to reach a lower level of the organization, i.e. level 1, by means of physiological processes not included in the original level, for example, by means of active transport across a membrane. In our field theory, an operator, called the S-propagator, represents the mechanism of transport of the interaction from one level to another. The transport takes place in three successive steps. First, the interaction is propagated within the source at r_i by means of the operator $\boldsymbol{P}_i$; next, the interaction is transformed by the structural discontinuity such that another functional interaction ϕ is propagated at a lower level; and finally, the interaction ψ is propagated at the higher level in the sink at r_0 by means of the operator $\boldsymbol{P}_0$. In mathematical terms, the S-propagator may therefore be written as: $\wp_{io} = P_0 \phi P_i$, which corresponds to the diagram:

$$r \xrightarrow[\wp_{io}]{\psi} r_0$$

The reasoning applied above for two hierarchical levels of a biological system is perfectly general. It may be extended to any number of levels of the structural and functional organizations since each unit at a given functional level belongs to a certain level of the structural hierarchy (Figure VI.5). Plate VI.1 illustrates the propagation of biological fields at several levels of a hierarchical organization. (In this connection we may turn

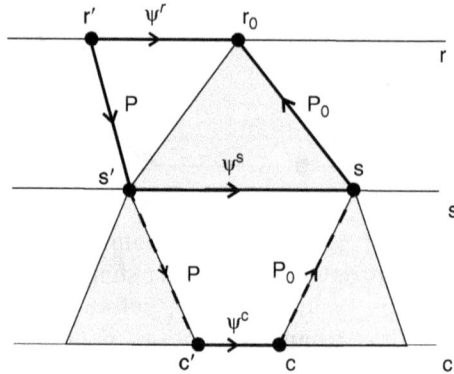

Figure VI.5: The interaction ψ^r in the r-space is the result of the interactions ψ^s and ψ^c at lower levels of structural organization, i.e. in the s-space and c-space, each space having a specific scale. For a given level of organization, the passage from one level to another takes place on a time scale that depends on the physiological function. Thus, the physiological function depends on a space scale as well as a time scale, as also shown in Plate VI.1.

to Figure IV.4, which shows the relationship between the structural and functional organizations of a biological system.) For example, one functional level of nervous activity, related to synaptic efficacy, involves structural units corresponding to *short-term synapses*, whereas another, related to transmission of potential, involves those corresponding to *long-term synapses*, which occupy a certain volume containing ion channels and macromolecules. Thus, the distinction between these two types of synapses is purely functional since the former ensure the transmission of membrane potential (on a time scale corresponding to the variation of membrane potential, i.e. several milliseconds), and the latter only modulate the transmission (on a time scale of the order of seconds). Synapses evidently constitute structural discontinuities and, in the case of nervous activity, the property of non-locality is linked to the existence of structural discontinuities separating well-defined volumes. We may therefore admit that, in a biological system, *the existence of a propagator is contingent upon the presence of a structural discontinuity.*

As illustrated in Plate VI.1 (in which Figure VI.5 is reproduced as an inset for convenience), the propagator allows the field to move across a level of functional organization *within* the structural hierarchy of a biological system. As in Figure IV.4, the x-axis corresponds to the space of structural units, the y-axis to the functional levels, and the z-axis to the structural levels of the biological organization of the system. The zigzag represents the coupling between the two functional hierarchical systems, i.e. it corresponds to a functional interaction between two units. The arrows on the inset represent the propagators according to the *connectivity* between the units at each level. The *density* of these arrows depends upon the density of the units at the lower level of organization. This is why we refer to the *density-connectivity function,* noted (π). The action of the propagator depends on the density of units (ρ). The inset in Plate VI.1 shows that, from a mathematical point of view, the propagator P results from the composition of two applications: $P[\psi'] = \mathbf{P_0}\psi^s\mathbf{P}$. This propagator can therefore be calculated. The inset also shows how the levels of structure for a given function intervene in the propagation along the ψ-axis, and how the levels of structure for another function intervene in the propagation along the μ-axis. Each of these functions involves structural units at different levels of the structural hierarchy as represented on the vertical axes in Plate VI.1.

The propagator P integrates the effects of several functions, such as the function μ (synaptic efficacy, in the case of the nervous system), as well as the function ψ (variation of membrane potential), which naturally contains the function μ. It thus enables us to describe *the action of one volume on another.* The field variable represents the "transport" of the interaction between two volumes in the space under the influence of the field, which exists at all points. If the relationship between the two volumes is *local,* as is generally the case in physics, then at a given point, the value of a field at a given instant depends on that of the immediately neighboring field at the immediately preceding instant. In contrast, if the relationship between the two volumes is *non-local,* as is the case in biological systems, the value of the field at a given instant depends on that of a field situated at a certain distance in space and time, and is determined by the field propagators. As we have seen, the property of *non-locality*, which is a consequence of hierarchical organization in a continuous medium, is an essential characteristic of all biological systems. Non-locality represents the effect of space on the dynamics of biological processes (D-FBS), just as the property of *non-symmetry* represents the effect of space on the organization of physiological functions (O-FBS). Figure VI.6 sums up these two fundamental aspects of a formal biological system.

Multiple time scales play a fundamental role in biological processes

In a biological system, the level of functional organization is defined by the global behavior of a set of structural units occupying a certain position in space. Acting as sources and sinks, these structural units emit and receive elementary functional interactions, each of which is represented by a field variable at each point in space at each instant of time. The field operator allows us to determine the value of the field variable,

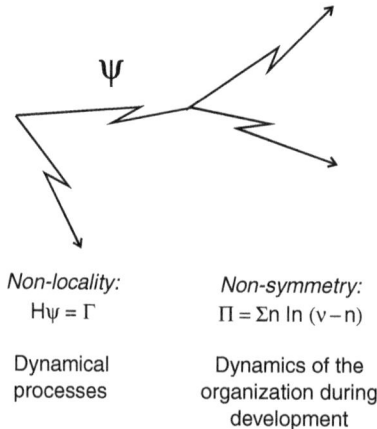

Non-locality:
$H\psi = \Gamma$

Dynamical
processes

Non-symmetry:
$\Pi = \Sigma n \ln (v-n)$

Dynamics of the
organization during
development

Figure VI.6: In a biological system, the D-FBS, i.e. the dynamics of biological processes, and the O-FBS, i.e. the organization of the physiological functions, are characterized respectively by the properties of *non-locality* and *non-symmetry*.

at the point at which the sink is located, from its value at the point of emission, at which the source is located. The distance between these two points is covered at a certain rate during a certain time. Thus, the abstract notions we have introduced actually correspond to the implicit physiological effect of time, i.e. to the time scale of the biological phenomenon at a given level of organization.

The global behavior of a set of units belonging to a particular level of biological organization over a period is provided by the solution of the field equation. The components of this equation consist of the phenomenological relationships between the field variables, which describe the elementary physiological mechanisms at the point of space considered. Thus, the solution of the field equation at each point of space varies more or less rapidly as a function of time. For example, a significant variation in neuronal activity may occur within a few milliseconds, whereas the modification of synaptic efficacy may appear only after several minutes. Again, the synthesis of an enzyme from a structural gene may require a much longer time than the chemical reaction catalyzed by the enzyme. The interval of time that is significant with respect to the phenomenon observed defines the *time scale* of the process. In fact, each physiological function, considered as the collective behavior of a set of structural units, runs on its specific time scale, in other words, each level of functional organization possesses its own time scale (Plate VI.1). The existence of specific time scales at each level of functional organization of a biological system leads to some remarkable consequences.

Might it not be preferable to use a representation of functional biological structures incorporating *discrete* dynamics? However, our representation of biological systems based on *continuous* dynamics offers at least two major advantages. The first is that the hierarchical representation used allows us to break down the global dynamics of the system into simpler dynamics, one for each level of organization, the dynamics being decoupled in time since each level is defined by a specific time scale; *this representation leads to a temporal order.* The second advantage is of a more fundamental nature since physical reality obliges us to consider that biological phenomena occur in continuous time and space. From this point of view, we may consider that *time is closely related to the functional organization, whereas space is directly linked to the structural organization of biological systems.* Space scales define the structural hierarchy of a living organism in terms of macromolecules, cells, tissues and organs. Thus, the notion of physical force is linked to space in the structural organization, whereas that of the functional interaction is linked to time in the functional organization of living organisms.

The propagation of fields in space depends on the connectivity between the structural units of a biological system

We may recall that the functional organization of a biological system was deduced from the hypothesis of self-association. This hypothesis reflects the vital condition that each structural unit of the system must receive the product it cannot elaborate from some other unit. When self-association of this type occurs, a functional interaction is established between the two units. This paradigm represents a natural mechanism that is essential for the organization of the structural units. In fact, the biological system can only exist — and be observable — if it is stable. The stability of the elements involved thus appears to be the causative mechanism of self-association. In other words, *in a biological system, the elementary mechanisms are organized such that the resulting physiological function is stable.*

The structural units of a biological system are distributed in a space that can be defined by a set of coordinates. This suggests the existence of a relationship between the *geometry* and the *topology* of the system. On one hand, the *geometry* represents the spatial distribution of the structural units, i.e. the *density* of the units. In mathematical terms, this involves the introduction of a spatial metric unit that allows us to determine the distance between any two points in this space. On the other hand, the *topology* represents the functional organization of the system, i.e. the functional interactions between its structural units, in other words, the *connectivity* between these units. *How do we actually determine the relationship between the geometry and the topology of a biological system?* Our method, schematically represented in Plate VI.1, is based on the consideration of discrete groups of countable structural units, which are geometrically localized and connected by functional interactions between the different structural levels of a biological system.

At a given hierarchical level, the number of functional organizations depends on the number of structural units in each group and the number of groups, i.e. the number of structural units, at the immediately higher level. The calculation therefore involves two levels of organization, the organization at the level of the groups, and the organization at the level of structural units within a given group. This method allows us to calculate the intra-group and inter-group potentials of organization. At the limit of continuity, we move from a space in which the points are isolated from each other to a space in which the points are distributed in a continuous manner. We then replace the set of points in a given volume by the density at a point in space, i.e. we replace the distribution of discrete structural units in the volume by the continuous density of structural units in space.

We thus obtain two mathematical functions that may be called the *density*, and the *density-connectivity* functions, respectively noted ρ and π. The former represents the density of the groups of structural units, whereas the latter corresponds to the density of the units in a group, localized at a point in space, connected to the other groups localized at other points in the same space. The potential of organization of the system, which expresses a multiplicative relationship between the number of sinks and the logarithm of the number of sources, evidently depends upon the density and the density-connectivity functions. As we shall see below, this leads to some interesting results concerning the behavior of a biological system when it is subjected to constraints.

The biological field theory presented above allows us to determine the field equation representing the dynamics of a biological system at a given level of organization. The equation, as we may recall, is written in the form: $\mathbf{H}\psi = \Gamma$. The first step is the choice of **H**, the non-local field operator representing the transport between two volumes. Then, at each "point"[48] of the space of structural units, which is distinct from ordinary physical space, the quantity ψ — which may represent an electric potential, a molecular or ionic concentration, a conductance, and so on — varies with time. Consequently, the operator **H**, which describes the unidirectional transport between two points situated at a finite distance, must be non-symmetric, local in time, and non-local in space. On one hand, the operator must include the infinitesimal variations occurring with respect to time and, on the other hand, it must contain a non-local and non-symmetric term representing the interaction, respectively taking into account the effects occurring at a distance and the exclusive effect produced by the source on the sink. Furthermore, it may be non-linear and depend on the field variable at each point in space. Finally, the source term Γ, which is at the origin of the phenomenon, describes local mechanisms, i.e. the local dynamics driving the biological processes; in other words, it describes the underlying phenomenology at the lower level of organization. The source, which is the point of *input* with respect to a given level, is the point of *output* with respect to the lower level of organization. The resolution of such field equations is rather difficult, even by numerical methods, particularly since the mathematical property of the inclusion of spaces of structural units from one level to another implies considerable constraints at the higher level due to the constraints linked to the solution of the field equation at the lower level of organization. This biological reality, which is rather difficult to imagine, is the consequence of the hierarchical functional organization of the biological system.

The non-local dynamics of the variables of interaction characterized by field propagators, which depend on the density of the structural units and the connectivity between them, can be superposed on the biological organization of sources and sinks, represented by a mathematical graph as described in Chapter V. Here we have the important relationship between the topology

---

[48]In what follows, we shall assume, unless specified otherwise, that each "point" refers to a space of structural units, and not to ordinary physical space. For simplicity of expression, we shall drop the inverted commas.

and the geometry of a biological system. Clearly, we will require a thorough knowledge of the phenomenology to be able to determine the variables, the non-local operators and the biological fields involved. Moreover, the nature of the variables and the hierarchy of the stacked spaces, in each of which the biological fields vary with time, are major factors contributing to the formidable complexity of the theory, the formalization and the description of biological processes. Unfortunately, it is quite unlikely that any simple representation will turn out to be suitable for biological phenomena. Finally, as in physics, the understanding of the living organism will probably be obtained only through a very high degree of abstraction.

The nervous system illustrates the role of space in biology

The physical concepts of space and time have always been so heavily charged with mysterious connotations that we may well wonder if these notions could be easily applied to biology, at least from the theoretical point of view. Nevertheless, the analysis of experimental findings clearly reveals the importance of space in the natural framework of functional biological organization in which vector fields describe the dynamics of the interactions. Although of an essentially physical nature, the concept of space, in association with that of time, allows us to define a minimum number of properties underlying a dynamic theory of physiological functions.

The field theory of physiological functions is based on three axioms. These axioms are derived from the fundamental properties governing the existence and the spatiotemporal variation of physiological fields. They are directly deduced from quantitative models[49] of biological processes, from the molecular to the organ level, in all the major systems of the higher organisms, i.e. the cardiovascular, renal, respiratory and metabolic systems, each of which is regulated by the nervous and the endocrine systems.

---

[49]For a detailed description, see Chauvet's *Theoretical Systems in Biology*, Pergamon, Oxford (1996).

VI

The *first axiom* concerns *the existence at each level of organization of an excitatory field propagated by an S-propagator.* This field describes how the dynamics of the biological system varies with time. The mathematical solution of the field equation representing the process corresponds to a dynamic state in the functional organization considered.

The *second axiom* defines the relationship between the process and its level of organization by means of a characteristic parameter, i.e. *the time scale specific to the level at which the functional process operates.* This axiom is of great importance since it implies the decomposition of dynamic systems at each level into several, simpler subsystems, in which each state variable at a given level serves as a parameter for the next higher level. There is no direct relationship between the *rank* of the level in the hierarchical organization and the value of the unit of time at this level, although it is the hierarchical organization that defines the time scale. The rank of the level of organization is determined by the inclusion of the spaces of structural units into the framework of a field theory. We may recall that Figure IV.4 shows the relationship between the levels of the structural organization, on one hand, and those of the functional organization of biological systems, on the other. The time scale, which is an essential characteristic of the physiological function, introduces a temporal order in the elaboration of a biological system. From an intuitive point of view, this temporal order is linked to the stability of the system. In other words, the biological system is functionally and hierarchically organized so as to be stable.

The *third axiom* bears on *the nature of the excitatory fields.* In most cases, the biological field is of the *excitatory-inhibitory* type. It is generally more convenient to consider the observed field as the resultant of two abstract fields with opposite actions. The field vector then possesses at least two components, one describing the activating effect, and the other the inhibiting effect. In fact, physiological systems are often regulated by two antagonistic actions, as in the case of the opposing effects of the insulin and glucagon hormonal systems in the regulation of glycemia, or the pairs of antagonistic muscles involved in the control of movement. And, as we shall see below, this also holds true for synapses in the nervous system.

Thus the excitatory field, which implies, on one hand, the existence of the functional interaction and, on the other, the finite rate of propagation of this interaction, reflects the principle of vital coherence. Among the numerous examples of excitatory fields that might be given, let us just mention the hormonal system from which the notion of the neurohormonal field may be readily deduced, and the nervous system, the detailed study of which has only just begun. However, since the neuronal system offers a remarkable illustration of physiological integration, we shall take just take a brief look at some fundamental aspects of nervous activity.

Let us now apply the biological field theory to nervous tissue (Chauvet, 2002). Nervous activity is fundamentally based on the effect produced by the transmission of an action potential by one neuron (the source) to another (the sink). The potential of the postsynaptic membrane is modified by local transformations due to synaptic transmission, i.e. the emission and diffusion of the neurotransmitter across the synaptic cleft, followed by its binding to the appropriate receptor, the opening of ion channels and, finally the flow of ions across the postsynaptic membrane leading to the modification of the local potential. The set of transformations that occur between the presynaptic and the postsynaptic membranes is represented by a *parameter* called the *synaptic efficacy*, which may be considered analogous to the property of *conductivity* in physics. The output, i.e. the local membrane potential, is equal to the product of the input, i.e. the action potential, and the conductivity. When the sum of all such local potentials is greater than a certain threshold value characteristic of the neuron, a new action potential will be emitted towards another target neuron, and so on.

How do we characterize the functional interaction between two neurons? This might be done, for example, by means of the frequency at which the action potentials are emitted. Neurophysiologists call this frequency the *activity* of the neuron since it corresponds to a measurable quantity that can be modified by the neuronal cell body. The local activity, observed at a point, must therefore satisfy a field equation for which the mathematical solution corresponds to the propagation of neuronal activity in time and in space. Furthermore, this activity must be observable by means of appropriate methods. In practice, however, a more convenient choice of field variable turns out to be the somatic potential, i.e. the membrane potential measured in the neighborhood of the cell body, or the axon hillock at which the axon extends from the neuronal cell body. The activity of the neuron may then be deduced from the number of action potentials emitted per unit time.

Finally, the behavior of a set of neurons, which is reflected by a global action, such as the peripheral nervous action on another set of neurons or on a set of muscles, and so on, defines a level of functional organization. This is illustrated schematically in Plate VI.1. The dynamics at a given level of organization operates on a specific time scale corresponding to that of action potentials, i.e. a time scale in milliseconds. In a continuous spatial representation, each neuron is associated to a point in the *space of neurons*. However, we have to take into account not only the density of neurons but also the connectivity between the neurons. This connectivity is established by the set of synapses existing on each neuronal cell membrane so that we have to consider the synaptic density at a point in the *space of synapses*, which corresponds in fact to a neuron. Thus, each point of space at the higher level, i.e. the space of neurons, is itself a space at the lower level, i.e. the space of synapses.

In terms of neuronal activity, a series of stimulations leads to the modification of the local postsynaptic potential, and the set of the local modifications leads to a global modification at the neuronal level. If this global modification exceeds a certain threshold value characteristic of the neuron, it will trigger an action potential corresponding to the neuronal activity. The functional interaction is transported from the presynaptic to the postsynaptic neuron more or less completely according to the value of the synaptic efficacy. From the physiological point of view, what is important here is that the synaptic efficacy, which is not a fixed parameter but a *state variable* of the system, determines the value of the local modification and, consequently, the activity of the neuron. In other words, the behavior of the set of synapses leads to the modification of the output, i.e. the neuronal activity. *The dynamics operating at the level of organization defined by the synapses is that of the temporal modulation of synaptic efficacy in the space of synapses; it therefore constitutes the second field variable.*

The structural units involved in the modification of synaptic efficacy are complex structures, comprising not only the synapses but also certain extra-synaptic and cytoplasmic elements, acting as long-term synapses, which we have called *synapsons* (a word constructed on the semantic model of certain particles in theoretical physics). Our biological field theory, applied to neuronal activity, predicts that *there must be a functional interaction between synapses.* This prediction is based on the non-local nature of the phenomena occurring at each level of organization. In effect, there is a structural discontinuity between two synapses represented by the extra-synaptic elements inside and outside the cell membrane, as well as the intracellular elements of the cytoskeleton and the extracellular medium, in which multiple sets of chemical reactions are known to occur. Thus, *the property of non-locality observable at the synaptic level allows us to predict the existence of heterosynaptic effects.* In other words, two neighboring synapses will affect each other even in the absence of any excitatory input.

Thus, the three axioms of the biological field theory, as enunciated above, are satisfied by nervous tissue since there are several levels of functional organization, each possessing an excitatory field. It has been shown that the synaptic modifications responsible for short and long term processes of memory operate on a much longer time scale (ranging from seconds to hours) than that of the variations of neuronal activity (occurring in a few milliseconds). This suggests that the parameter of the coupling between the two levels of organization corresponds to time. Again, the existence of the activatory and inhibitory synapses, linked to the molecular nature of neurotransmitters, leads to the introduction of an observable synaptic field and, as a consequence, a *non-observable* field of activatory and inhibitory activities.

The nervous system may therefore be considered to possess at least two levels of functional organization, i.e. the neuronal level at which the function of the set of neurons corresponds to the activity of a neuronal network, and the synaptic level at which the function of the set of synapses corresponds to the emission of an action potential. This description of the nature of the dynamical aspects of a biological system (D-FBS) coherently reinforces that of the structural aspects (O-FBS) as previously described, thus leading to the functional organization presented in Chapter II. The relationship between the D-FBS and the O-FBS, i.e. between the geometry and the topology of the functional organization of a biological system, generalizes the widely accepted connectional theory of artificial neuronal networks to *real* neuronal networks.

As we have seen, the density and the density-connectivity functions link the geometry and the topology of a neuronal network. These spatial functions take into account the number of structural units, neurons or synapses, in an infinitesimal elementary volume, implicitly including the functional organization of the group of neurons or the group of synapses, in particular through the function of connectivity or the probability of the connection between the structural units. However, *these geometrical functions may also act as variables with respect to time*, as actually observed, for example, during the development of a biological organism or during post-traumatic reconstruction. *The connectivity that subtends the propagation of fields* thus appears as the basic link between the time-variation of the graph of the biological system (Chapter V) and the dynamics of the processes of the system as described above. The property of connectivity is a mathematical function that is essential to *a biological field theory, which is of a topological and geometrical nature, whereas a physical field theory is uniquely of a geometrical nature.*

Dynamic processes in biological systems and mathematical graphs are related through topology and geometry

One important consequence of the coupling between the topology and the geometry of a biological system, i.e. the D-FBS and the O-FBS of the living organism, is that it determines the selection of structural units operating in specific groups. The development of a biological system raises an important question. How does the biological system develop while the field dynamics, which describe the *action* of one structural unit on another, and the topology, which characterizes the *organization* of the system through the connectivity between these units, vary simultaneously in time and space? This is the case, for example, in a group of neurons during the development of the nervous system or, more generally, in any group of cells in the process of differentiation during which the multiplication of cells corresponds to their specialization in a particular physiological function. Let us now see how this problem may be formulated in mathematical terms.

As a physiological process develops, the number of cells, or the density of cells per unit volume, varies. As we have seen, a biological field may be used to represent this process. Moreover, as cell densities vary in space and in time, they may be represented by a variable field under the influence of a field operator. And since each physiological function satisfies a field equation, the formalism of the field theory may be used to describe the developmental process. In general, field operators depend on the physiological processes, and cell densities depend on the lower levels of structural organization as well as the density-connectivity functions at the adjacent levels that determine the couplings between the different levels (Plate VI.1).

The total number of structural units in a given space may then be obtained by mathematical integration. However, since the hierarchical nature of the biological system implies that a point at a given structural level is actually a space at the immediately lower level, the calculation of the number of structural units involves the density and the density-connectivity functions associated with each successive pair of organizational levels. In a way, the spaces associated with each structural unit are stacked rather like Russian dolls. Let us now suppose that the system is under a global organizational constraint, such as that of a constant density of structural units at the lowest level of the system. The field equations show that a time-variation of the density-connectivity functions leads to a redistribution of structural units. Furthermore, calculations show that the potential of organization of the system, initially considered constant, actually varies with time. This property may be expressed as follows:

> Following a time-variation of the density-connectivity functions, a biological system subjected to the condition of a constant global organization, the dynamics of which are described by the functional interactions and the densities of structural units at each level, will have a time-dependent potential of organization, and will undergo a redistribution of structural units.

This property is of considerable importance since it establishes an essential relationship between all the concepts introduced in the preceding chapters. In particular, it expresses the coupling between the geometry of the system (described by the density functions) and its topology (described by the density-connectivity functions). It also establishes the abstract relationship between the dual representations discussed in Chapter IV, i.e. the (ψ, ρ) representation, corresponding to the functional interactions, and the (N, a) representation, corresponding to the occupation numbers of the classes of theoretical biological systems.

In other words, the field equations describe the coupled time-variation of the physiological processes and the organization of a biological system, i.e. the reciprocal influence of the dynamics of the functional interactions and the organization of the system operating under certain constraints. These constraints may correspond, for example, to a law of conservation of the number of structural units, or a law of conservation of the number of functional interactions in a given space. Numerical computer simulations have demonstrated that such constraints actually produce a selective spatiotemporal redistribution of the density of structural units in a biological system.

The property stated above leads to several other predictions that would surely be interesting to test. For example, *during the development of the central nervous system, we may expect groups of neurons to be selected to perform specific functions, with a consequent redistribution of the associated synapses.*

Similarly, we may expect groups of antibody-producing cells in the immune system to undergo selection and redistribution during the development of the organism.

Another important consequence of the coupling between the topology and the geometry of biological systems, i.e. the D-FBS and the O-FBS, is that it describes the process of self-organization of a living organism. We must admit that certain terms used in biology, such as *self-organization, complexity* or *autonomy,* have a somewhat mysterious connotation. This is probably because it is rather delicate to define these specifically biological concepts, even though they may be commonly used. Thus, some physiologists might find it difficult to accept a definition of self-organization based essentially on the structural aspects of a biological system. However, several scientists — other than physiologists — have dealt with the problem of self-organization in purely structural terms, using a highly figurative, symbolic formulation. Thus, we have Schrödinger's principle of order arising from pre-existing order, or Von Foerster's hypothesis of order emerging from random signals, both ideas being based on information theory, or again Prigogine's concept of order stemming from thermodynamic fluctuations, based on the theory of irreversible processes. It is therefore interesting to see how the ideas of self-organization, complexity and autonomy may be defined within the framework of our functional biological theory. The formulation in terms of functional interactions is particularly useful since it gives these concepts a mathematical reality that is relatively easy to interpret in biological terms. Moreover, it allows us to formulate the principle of *functional biological order emanating from a hierarchical organization,* and to demonstrate that the functional hierarchical organization of a biological system leads to a particular mode of development, corresponding precisely to the *self-organization* of the living organism.

Let us first consider the difference between *organized* and *self-organized* systems. We may say that a system is organized if its internal state varies under the influence of an external cause, and ceases to do so when the cause is suppressed. For example, the structure of a metal rod corresponds to an arrangement of elements that are completely stabilized by physical interactions. If the rod is twisted, the set of forces acting between its elements will be perturbed, leading to a modification of its structure. Again, if the forces acting between the elements are modified, the rod will change its shape. This type of definition, which expresses the relationship between external causes and internal effects, can be readily expressed in mathematical terms. It may also be extended to a self-organizing system when the organizing forces are *internal* with respect to the system. This is the case, for example, in a society that organizes itself under a system of laws, i.e. a society that establishes its own interactions. Obviously, the mathematical formulation of self-organizational processes will then be identical to that of an organizational process, the only difference being in the nature of the applied forces, which are internal in the former and external in the latter. It is equally obvious that the internal forces change with the system such that there is a reciprocal action between cause and effect. Thus, it is easy to see how a material system subjected to internal forces, such as shear forces, may be deformed and change from one state to another with a different molecular configuration.

However, can such concepts of organized and self-organized systems be applied to living organisms? We may recall that living organisms can be represented by two abstract systems, i.e. the O-FBS describing the topology of the functional interactions, and the D-FBS, expressing the dynamics of the physiological processes. These systems are coupled such that each functional organization at an instant t, represented by the O-FBS(t), is associated with a certain dynamics of the D-FBS (Figure VI.7). Thus, *the property of self-organization will be coherently expressed if the biological system can be shown to move from one stable state of the dynamic system to another when it is subjected to a modification of the topological system.* In other words, a living organism may be considered to be self-organizing if a modification of its topology, i.e. its functional interactions, for example, by means of a suppression of

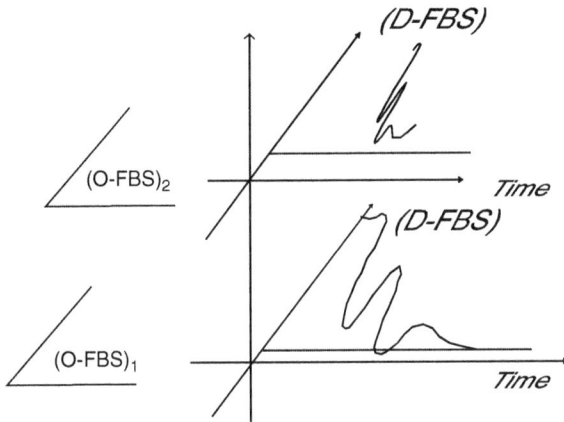

Figure VI.7: The topology and the dynamics of a biological system are coupled so that the dynamic system (D-FBS) is modified when the organization of the system (O-FBS) undergoes a change. Each O-FBS(t) has its corresponding D-FBS.

sources or sinks, causes it to move from a state in which the dynamics of its physiological processes are stable to another in which its dynamics are again stable. In the language of modern mathematics, the property of self-organization of a biological system implies that a variation of its topology leads the system from one stable attractor to another, i.e. from one region of stability to another in phase space. In Plate VI.2, the schema (c) illustrates this definition of self-organization in the phase space of the dynamics. The vertical axis represents the functional organization of the biological system at various instants, and the other two axes represent the corresponding dynamics of the system in the phase space. Then, according to our definition, if there is a stable attractor in each plane, the biological system possesses the property of self-organization.

Plate VI.2 illustrates the property of biological self-organization. The endocrine system corresponding to the hypothalamo-hypophyso-testicular axis is shown in (a). This portion of the endocrine system regulates the pulsating secretion of male sexual hormones in the organism. These hormones include the luteinizing hormone (LH), noted $L(t)$; the gonadotropin releasing hormone (GnRH), noted $R(t)$, both of which are secreted by the hypothalamus; and testosterone, noted $T(t)$, secreted by the testes. The corresponding cybernetic schema is shown in (b). Experiments suggest that the secretion of testosterone, which is independent of LH, is controlled by the adrenal cortex. For this functional organization, noted (O-FBS)$_1$, the dynamics of the L-R-T system are represented by curves, in which the initial state is situated inside a closed domain. As the time t changes, we have a limit cycle in the phase space (L,R). The limit cycle, to which all the trajectories starting from the initial state tend, is shown in (c). The new functional organization, noted (O-FBS)$_2$, produced, for example, by the chemical castration used in the treatment of a tumor, and the new limit cycle attractor are shown in (d).

In the regulation of male sexual hormones by the hypothalamus, the hypophysis and the testes, i.e. the (O-FBS)$_1$, the castration of the testes leads to a new organization, i.e. the (O-FBS)$_2$, while the system, shifting from one attractor to another, remains stable in spite of the castration. This stability implies the existence of the property of self-organization in the biological system. The important point here is that *the notion of the "internal force" concerns the topology of the biological system so that a particular cause, such as a micro-mutation or some other structural modification, will affect the dynamics of its physiological processes.* Conversely, the effects of the dynamics of physiological processes on the topology of a biological system could explain the evolution of new species. According to this theory, the principle of vital coherence, which expresses physiological stability, implies the property of self-organization in living organisms.

Thus, the mathematical formulation of a physiological phenomenon in terms of graphs and a biological field theory leads to interesting conclusions. In particular, a field equation can be associated with each level of organization of a biological system, considered as a set of hierarchical functional interactions. Each level of organization operates on a specific time scale so that the perturbation of a variable at one level will affect the dynamics of the system at another level, since the variable at one level acts as a parameter at the other. *This corresponds, in a sense, to the "transport" of fluctuations, from one level to another, within the biological system, which will then react by a process of self-organization.* Pathological variations corresponding to such fluctuations are commonly observed in living organisms and if the affected biological system cannot reorganize itself according to its internal constraints, it will not be able to survive.

Of the three mysterious concepts mentioned above, i.e. self-organization, complexity and autonomy, we have dealt with the first in some detail. Let us now briefly consider the other two. Can the *complexity* of a biological system actually be measured? We may recall that the functional organization represented by a hierarchical graph corresponds to the dynamic process described by the field equation. Clearly, the complexity of the vectorial field equation will increase with the number of functional interactions involved and the number of couplings established between the different levels of organization. This corresponds to an increase in the complexity of the topology of the system. There is clearly a direct relationship between the functional complexity and the value of the potential of organization as defined above. Thus, *the functional complexity of a biological system may be measured by means of its potential of organization.*

The biological concept of *autonomy* has raised much controversy, mainly because of the fundamental difficulty of defining a living organism. According to our definition, a biological system is composed of structural units that interact functionally. Thus, if the units were completely independent, each would be the seat of the same dynamic process as all the others, so that there would be no difference between the global behavior of the system and the individual behavior of any of the units. However, the description of a biological system in terms of its hierarchical organization excludes such functional independence between its structural units. There is an essential difference between the organizational level of a living organism and that of the environment. Thus, although the structural units may be physiologically independent of each other, each unit will be physiologically dependent on its environment. In this case, of course, the dependence is not of a *functional* nature. We may therefore consider that *a biological system has the property of autonomy if there is no higher organizational level of integration*, in other words, if it is functionally independent of all other sets of structural units. In general, such a biological system is dependent only on the environment. From this point of view, the sexual function, common to all higher organisms, is certainly a factor that is not in favor of the autonomy of the individual organism. But here, we come to the frontier between physiology, the science concerned with the functions of living organisms, and sociology, the science that deals with the behavior of groups of organisms. However, crossing this frontier is far beyond the scope of the present work.

Can mathematics solve the mystery of the evolution of the species?

As we have seen, the coupling between the mathematical graph of the organization of a physiological function and the corresponding dynamic process leads to two essential concepts: the selection of structural units that operate as coordinated groups and the self-organization of the biological system. Of course, these notions are closely linked, but what precisely is the nature of the dependence? It is obvious that biological systems have different degrees of complexity and autonomy. Living organisms obtain their energy from the environment in very different ways. Thus, from a geographical point of view, an animal is obviously more autonomous than a plant, but with respect to essential nutritive needs it is the plant that is more autonomous than the animal. Now, which of these organisms has the greater autonomy? Which is the more complex? Which is better organized? The mathematical definitions given above allow us to demonstrate certain properties. For example, *the autonomy of a biological organism generally increases with its potential of organization*. However, this does not tell us whether it is the increase in autonomy that leads to the increase in the potential of organization, or inversely, whether an internal reorganization of the topology of the system (O-FBS) follows the fluctuations in the neighborhood of the stationary states of the dynamics of the system (D-FBS). In the latter case, the dynamic process would tend towards a completely different stable stationary state associated with the establishment of some vital parameter (for example, the synthesis of a molecule indispensable to the biological system), thus increasing the autonomy of the organism. It may well be that both these processes occurred in the course of evolution. How are we to find out if this is true? The precise knowledge of the functional organization of the different species and the associated dynamic physiological processes may be expected to provide the answer. What is certain is that the evolution of life from the simplest amoeba to the most highly developed mammal is directly associated with an increase in biological organization, in other words, an increase in functional order. Nevertheless, can it be held that an amoeba is less autonomous and less complex than a mammal?

Clearly, these questions will find an answer only when *the* complete *map of the functional organization for each living species can be drawn.* This gigantic project, when finally achieved, would surely solve the riddle of evolution. Unfortunately, the technology required for the construction of functional maps of living organisms has yet to be fully developed. In the meantime, it would seem worthwhile to seek, as we have tried to do above, the explicit relationships between fundamental concepts such as the self-organization, complexity and autonomy of living organisms. Even here, the difficulty is considerable since, for example, the comparison of various species with respect to their autonomy would imply a thorough knowledge of the dependence of each species on its environment. While it is relatively easy to define the nutritive elements that are essential for the life of an organism in its environment, it is rather more difficult to determine the form under which such substances are actually assimilated by the organism.

VI

Are physical structures and physiological functions complementary aspects of the Universe?

The abstract vision of a living organism as a dynamic system of physiological processes associated with a hierarchical topologic system leads to a set of field equations governing the functional interactions between the structural units of the organism. Thus, the geometrical theory that applies to inert matter in physics has its counterpart in the geometrical and topological theory that applies to living organisms. The integrated physiology of a biological system may then be conceived in terms of the dynamics of a set of functional interactions described by fields under the influence of propagators at each level of hierarchical organization. The existence of activatory-inhibitory fields and feedback loops in biological systems produces oscillating fields or asymptotic behaviors that define the time scales at each level of organization. In addition to this very specific aspect of the biological universe in which time appears to be an intrinsic reality, it is interesting to note the existence of opposed structural and functional realities in physics and biology, i.e. that of matter and anti-matter in the former, and that of activation and inhibition in the latter. The profound significance of this comparison, reflected by the field theory in physics and our multiple-field theory in biology, lies in the fundamental principles of regulation, which are expressed by the laws of conservation of matter in physics, and by the principle of vital coherence, or the conservation of life, in biology. Finally, we may well wonder whether the existence of life, intimately linked to physical matter, does not correspond to a cosmologic reality of the Universe, which would appear to be composed of two worlds of energetically opposed structures, and two worlds of functionally antagonistic organizations.

Epilog

The Mathematical Nature of the Living World

Epilog

THE MATHEMATICAL NATURE OF THE LIVING WORLD

"The day we succeed in understanding life as a function of non-living matter, we shall discover that it possesses properties far removed from those we thought it possessed."

Claude Lévy-Strauss, La pensée sauvage (1962)

The aim of theoretical biology is to discover the fundamental unity of a living organism and, by extension, the greater unity of life in the universe. By analogy with physics, a unifying theory in biology may be expected to depend on a limited number of general principles governing the origin of life in non-living matter. However, it would be foolhardy to underestimate the formidable difficulties involved in the determination of such principles. We have therefore set ourselves the far less ambitious task of constructing a formal representation of the functioning of a living organism. This approach to integrative physiology should allow the interpretation of certain biological phenomena within a coherent theoretical framework.

In this work, we have tried to present the unity between biology — or more precisely, physiology — and physics; in other words, between life and non-living matter, in terms of dynamic processes. It would of course be perfectly illusory to believe that a problem as complex as this could be treated in any simple way. Nevertheless, the approach we have used should be reasonably accessible since the theory we propose explains the organization of a biological system essentially in terms of functional interactions between its structural units. Apart from the complex mathematics underlying the theory, all the concepts involved are fairly straightforward. Thus, it will be readily acknowledged that the biological universe is constructed according to the geometry of the physical universe, and that the living organism is characterized by its topology and the optimum principles governing the irreversible dynamics of the organization and the physiological processes of the biological system, in other words, the mechanisms of life.

Of course, all these ideas will need to be experimentally tested. Given the enormous amount of information that has been acquired in various fields of biology, the hypotheses put forward can only be justified by integrating all the experimental results within an appropriate theoretical framework. Of course, if this integration proves impossible, then the theory will have to be modified, or perhaps even abandoned. As the reader must have noticed, the terms "biology" and "physiology" have often been used interchangeably. We make no apologies for this since, within the framework of our theory, the sharp distinction drawn between biological structure and function literally transforms biological science into physiology. The difference between living and non-living matter is such that biology, the science of life, is clearly seen as the science of functional processes, i.e. physiology. Obviously, the study of the material structures that support the functional processes is necessary — but not in itself sufficient.

To be sure, most biologists, more immediately concerned with biological structures than physiological processes, will probably reject the idea of a general theory based on optimum principles. How could anyone pretend to capture the infinite diversity of the living world within the framework of a universal biological theory? In the present state of our knowledge, this would certainly seem an impossible task. However, our objective here is of a totally different nature. It can hardly be denied that the physiological processes operating in organisms ranging from the simplest unicellular organism to the most complex mammal are rather similar. Both consume oxygen, synthesize molecules essential to their existence by breaking down molecules taken up from the environment, discharge waste matter, and so on. In addition, the learning mechanisms in all the organisms with a central nervous system are identical. We have therefore tried to construct an abstract framework that would allow us to determine certain basic principles underlying the functioning of living organisms or, more precisely, to identity the elements that might be common to all fundamental biological mechanisms.

Advances in the scientific explanation of nature have often been stimulated by the discovery of identical mechanisms underlying what appear to be, at first sight, very dissimilar phenomena. For example, when Evariste Galois (1811–1832) elaborated the mathematical theory of groups, he could hardly have imagined that his discovery would affect all fields of science. The powerful mathematical techniques derived from group theory have been applied to sets as different as those of real numbers and geometrical transforms. In physics, the analysis of groups of symmetry allows the identification of absorption lines in the spectra of molecular vibrations. It is surely fascinating that the energy levels of a molecule can actually be determined by applying group theory to its properties of symmetry. Group theory is extensively used today in such apparently unrelated domains as nuclear physics and linguistics, to mention but two. Our work has been directed to adapting the concepts of the theory of physical fields to key biological phenomena in the hope of bringing out the fundamental differences between living organisms and non-living matter.

Is the time now ripe for a unifying theory in biology?

The description of phenomena at the molecular and cellular levels in living organisms reveals the existence of a fundamental biological unity at the gene level. This is not surprising since the chemical processes occurring in the cells are all directly controlled by gene dynamics. Biological unity is reflected in the increasing use by geneticists of a terminology that is analogous to that used by physicists. Thus, some of the terms used with reference to nuclear particles, such as protons, nucleons, bosons, and so on, are semantically echoed by biological terms, such as mutons, cistrons, operons, and so on.

The fundamental difference here is that the physical terms describe *structural* units of non-living matter, whereas the biological terms refer to *functional* units of gene activity. For example, a gene mutation can lead to a specific loss of function due to the modification of the structure of a protein. Thus, the functional unit involved in the basic gene mutation has been called the *muton*. In fact, the muton is a nucleotide pair within a *cistron*, which itself is the smallest alterable element of the one-dimensional structure of genetic material.

While biological unity at the gene level is fairly well established, it is rather more difficult to determine biological unity at the organ level of a living organism. Nevertheless, it is possible to isolate certain theoretical subsystems of the organism. One example is given by the kidney, which ensures the cleansing of blood and the maintenance of the acid-base and hydroelectrolytic equilibria of the body. The kidney can be considered as being made up of a large number of functional units, each comprising a *nephron*, together with a portion of the *interstitium* and the *vasa recta*. Each renal unit possesses its own entry and exit, thus communicating with the environment and ensuring exchanges. The physiological function of the set of renal units defines a certain level of functional organization with properties corresponding to a structural unit at a higher level of organization. Another example is provided by the lung, in which the terminal bronchiolus, or *alveolus*, plays the role of the respiratory functional unit, ensuring ventilation, gaseous exchange and capillary perfusion. Again, in the cerebellum, the so-called "*microzone*", recently identified as a set of neurons that activates an element of a group of muscles engaged in a particular movement, might be considered as the functional motor unit of the specific muscular activity. Unfortunately, it is still rather difficult to define the functional units involved in the cardiovascular system, the digestive system, the muscular system, and so on. The identification of the functional units in all the organs of the body presents enormous difficulties, especially in the absence of a general theoretical framework. This is why we believe it is important to lay down the bases of a possible biological unity in living organisms.

Unifying theories have always fascinated the human mind, and this must surely have stimulated the search for what Einstein described as the "simple" image of the external world. According to this vision, the material universe must satisfy a very restricted number of laws derived from a great unifying theory. Such a theory could conceivably reconcile the old Platonic idea of the universe with that of modern physics. However, the unity that emerges in biology is very different from that described by physicists. This disconcerting aspect of nature constitutes the profound mystery of life. Whereas it is relatively easy to accept the idea of a physical universe that may be encompassed by a single theory that remains to be formulated, how do we explain the functioning of a living organism composed of a multitude of different functional systems, closely interconnected and regulated, each of which is a hierarchical system obeying laws of its own? This diversity of functional systems corresponds to a multiplicity of formalisms each of which requires specific models to describe the various functional systems. A unifying theory in biology must therefore be situated on a much higher plane, at a level of far greater abstraction and integration. If such a theory is ever constructed, it will incorporate the unity of the living organism by means of the coherent integration of all its physiological mechanisms, taking into account the different levels of organization and the specific aspects of biological systems. This, of course, is precisely the final goal of integrative physiology.

What is the fundamental nature of non-living matter?

The electromagnetic field, which is characterized by the physical laws of structure given by Maxwell's field equations, corresponds to the interaction between two charges. The gravitational field is similarly characterized by the physical laws of structure given by the equations of Einstein's general theory of relativity. Although Newton's theory of gravity contains no field equations, Newtonian mechanics describes the interaction between two masses. Charge and mass are physical characteristics of matter, and the concepts of the electromagnetic and gravitational fields clearly bring out *the intimate relationship between the field, the geometry of space and matter.* In a sense, we may perceive the physical universe as a space containing islands of matter bathed by a field. Under certain conditions, the field may manifest itself in the form of waves propagated in space, such as those found when electromagnetic waves are emitted by an oscillating source. We may therefore expect to observe a similar phenomenon in the case of the gravitational field. In other words, an oscillating mass should produce *gravitational waves.*

However, the existence of gravitational waves, predicted by Einstein's general theory of relativity published in 1915, remained unproved for several decades, mainly because of the enormous difficulty of the experimental work involved. It was only when the Hulse–Taylor binary pulsar (PSR 1913 + 16) was discovered in 1974 that gravitational waves were finally detected. A binary pulsar comprises a pair of rapidly rotating neutron stars, one of which is a pulsar, orbiting each other. This binary pulsar, which emits regular electromagnetic pulses every 59.0299952709 milliseconds, acts as a high-precision cosmic clock. This is of interest because the space-time curvature inside and in the neighborhood of the binary pulsar is large enough to reveal the effects due to the very intense gravitational field and, in particular, the propagation of the gravitational interaction at a finite velocity. The general theory of relativity indicates that the dynamics of the binary pulsar must include additional forces of gravitational interaction due to the relative velocity of the two bodies. In turn, these additional forces modify the acceleration of the orbital movement of the binary pulsar. Thus, by determining the orbital movement with sufficient precision, it is possible to compare the theoretical consequences of general relativity with observations. In effect, the calculation predicts an acceleratory effect, corresponding to the reduction of the orbital time of the double-pulsar, which coincides with the observed value to within 4%. This surprisingly precise result demonstrates the value of the theory of general relativity in the interpretation of strong gravitational fields existing around each body, in the propagation of the gravitational interaction between two bodies at the speed of light, as well as in the important non-linear effects occurring in the region of weak gravitational fields between two bodies. The calculation of the velocity of gravitational waves gives the velocity of light with a precision of 1%. Thus, within the framework of the theory of general relativity, it would seem reasonable to consider *gravitational waves as propagations of distortions of the geometry of space and time.*

Now, is there any real difference between matter and its associated field? Since the field has its origin in matter, it may be considered as an extension of the source, possessing a material reality differing from that of the source essentially by its energy content. Thus, we may imagine an island of matter, containing an enormous amount of energy, as indicated by Einstein's equation: $E = mc^2$, surrounded by an extremely weak field. The equivalence between mass and energy would then imply that mass is distributed in totally different ways between matter and its field, i.e. very strongly concentrated in the former but very weakly in the latter. However, the equations describing the reality of matter and its associated field show that the laws of physical structure are quite different in the two cases. If the physicist's dream of a unique set of equations encompassing the structure of matter as well as the structure of the field ever comes true, then matter could be considered as a region in space where the field is extremely intense, and a body in motion would correspond to a field variable in time. The geometry of the universe, which is composed of the distribution of masses, would then correspond to the structure of the field. Current efforts in modern physics are directed towards the search for a unifying theory for the four major types of physical force, i.e. the gravitational and electromagnetic forces, and the strong and weak interactions of matter. If Hawkings' theory of superstrings ever achieves this objective, it would finally explain the origin of the universe in terms of a single original force. Human thinking has always been peculiarly attracted by explanations based on the unique cause of observed events since this would correspond to the fundamental simplicity of the universe, in Einstein's sense of the term, implying the *real* existence of the cause.

Epilog

Modern theoretical physics is based, on one hand, on *the theory of general relativity*, which takes into account the force of gravity in the space-time continuum, and, on the other, *the theory of quantum mechanics*, which takes into account the three other forces of nature, i.e. those of electromagnetism, strong interactions and weak interactions. The theory of general relativity gives us an insight into phenomena occurring on a cosmic scale, whereas the theory of quantum mechanics describes events on the atomic and the subatomic scales. Thus, *the search for a unified theory in physics amounts to the search for a quantum theory of gravitation* involving *the quantification of fields immerged in a curved space-time continuum.* The theory of superstrings will therefore have to include the concepts of gravitation. Ordinary three-dimensional space, as perceived by our senses, is of course the space in which Newtonian dynamics is valid. The explanation of gravitational interactions, in the framework of the theory of general relativity, requires a four-dimensional space-time structure. Finally, if electromagnetic interactions together with the strong and weak interactions of matter are included, the theory of superstrings will have to describe a *ten-dimensional universe.* This surely constitutes a tremendous theoretical challenge to the human mind!

What is the precise nature of the force acting between two particles of matter? Quantum theory offers an interesting answer. At an elementary level, the force corresponds to the interaction between two particles of matter. According to quantum theory, forces are supposed to be transported by *virtual, force-carrying particles* that differ from the other particles of matter by their *spin.* Each particle has a specific spin value that characterizes an invariant aspect of its behavior about an axis. For example, spin 0 corresponds to no particular orientation, spin 1 to a full turn (360°), spin 2 to a half-turn (180°), and spin 1/2 to two turns of a particle on its axis. All matter in the universe is composed of real particles with spin 1/2, whereas the force-carrying particles have spins 0, 1 or 2. A fundamental principle of quantum mechanics, the *Pauli exclusion principle,* states that two elementary particles of matter (spin 1/2) in an atom cannot exist simultaneously in the same quantum state. However, this principle does not apply to the force-carrying particles (spin 0, 1 or 2) so that these particles may be exchanged in numbers great enough to produce a strong interaction. Although the force-carrying particles really exist, they cannot be observed directly, which is why they are called *virtual particles*. The force acting between two particles, A and B, may now be expressed as follows: an elementary particle of matter A (spin 1/2) emits a virtual particle (spin 0, 1 or 2) at a velocity modified by the recoil of A due to the emission. The virtual particle then collides with another elementary particle B (spin 1/2) that absorbs it, the collision modifying the velocity of particle B. Finally, it is the transfer of this modification of velocity that corresponds to the force acting between the two particles A and B.

The gravitational force between two bodies is attributed to the exchange of virtual particles with spin 2, called *gravitons*. As in the case of the other particles in quantum mechanics, gravitons are associated with waves, called *gravitational waves*. In a similar manner, the electromagnetic force between two bodies is due to the transfer of virtual particles with spin 1, called *photons*. Thus, there is an exchange of virtual particles between the negatively charged electrons and the positively charged nucleus of an atom. The resulting electromagnetic attraction is very similar to gravitational attraction so that the electrons spin around the nucleus rather like the planets revolving about the sun in their elliptic orbits. Photons are therefore associated with *electromagnetic waves*.

Thus, the gravitational and the electromagnetic forces, which correspond in fact to the fundamental interactions leading to the structural stability of matter, are transported by means of an exchange of virtual particles. The *force field* therefore corresponds to a material reality reflecting the exchange of particles at each point of space.

What is the fundamental nature of life?

Given the present state of our knowledge, inquiring into the fundamental nature of life would surely seem to be a risky endeavor. Nevertheless, let us approach the problem through the functional organization of living organisms as described by the combination of elementary functional interactions between sources and sinks. Thus expressed, the biological universe is clearly distinguishable from the physical universe in which the structural organization is deduced from the combination of forces acting on elementary physical structures. As we have seen, biological systems are characterized not only by the hierarchy of structures but also by the hierarchy of functions and the hierarchy of regulations tending to the optimum control of the systems. The elementary functional interaction possesses two fundamental properties, i.e. the non-symmetry of the action from source to sink, which implies a local transformation in the sink, and the non-locality of the action in space, which arises from the hierarchical structure. These specifically biological concepts form the basis of our theory of integrative physiology.

In short, the functional biological interaction corresponds to the non-symmetric and non-local action of one element of matter on another. However, each element of matter — or structural unit — is itself a hierarchical system composed of other structural units, somewhat like a series of nesting Russian dolls. Thus, the organs of a biological organism consist of tissues, which are made up of cells, which in turn are composed of organelles, and so on, down to the smallest elements of matter. Finally, each physiological function of the living organism corresponds to the dynamics of a particular set of elements of non-living matter so that *life can exist in non-living matter only under certain conditions that are intimately linked to the properties of stability of biological systems.*

The first condition of life in non-living matter is the existence of functional biological interactions. This condition is satisfied by the *hypothesis of self-association,* according to which an element of matter that no longer performs a necessary elementary physiological function must receive the product of this function from some other element of matter that still possesses the function. The association of two structural units that are identical, except for a physiological function that may be absent in one and present in the other, occurs during the development of an organism thereby increasing the domain of stability of the dynamic functional system in spite of the increase in complexity this entails. Whereas the action of physical forces, which are symmetric in nature, leads to structural stability in non-living matter, the process of self-association leads to an increase in functional order in biological systems. *Thus, the stability of the functional dynamics of a biological system is the first characteristic of life in non-living matter.*

The second condition of life in non-living matter is that the organization of the functional interactions of a biological system must satisfy certain criteria of stability. But what exactly is the nature of this organization? The stabilizing, non-symmetric interactions of the functional dynamics of the biological system, i.e. the D-FBS, define the topology of the system as described by an oriented graph corresponding to the O-FBS. *The potential of functional organization of a living organism during its development* can then be determined by combinatorial mathematics. These calculations are based on the fundamental property of living organisms, i.e. *self-reproduction.* The highest value of the potential of functional organization characterizes a particular class of systems that are, by definition, *observable biological systems.* Since the biological system in the course of its development possesses the property of stability related to the monotonous variation of the number of sinks, the extremum hypothesis of self-association may be experimentally tested, thus providing a means of validating the theory of integrative physiology.

> The potential of functional organization of the biological system, written $\Pi = \Sigma\rho \ln (\upsilon - \rho)$ (where ρ is the density of structural units corresponding to sinks, and υ is the total number of structural units in the system), is *non-symmetric* since it concerns functional interactions that are by nature non-symmetric. In 1887, Boltzmann described the behavior of inert matter by means of a statistical expression, $H = \Sigma\rho\ln\rho$, a *symmetric* function in which ρ represents the density of states of matter in phase space. Inert matter, the substance of all living matter, is governed by the thermodynamic state function of *entropy* deduced from H (according to $\Delta S = -kH$) such that $\Delta S \geq 0$. In contrast, the organization of the physiological functions of

Epilog

living matter is governed by the biological state function of *orgatropy* deduced from the potential of functional organization Π (according to $F = \Pi_{max}$o *h*) such that $\delta F \leq 0$. The definition of orgatropy (Chapter V) clearly brings out the distinction between living and non-living matter. Since functional interactions contribute to the stability of the dynamics of a biological system, the presence of life in non-living matter during the development of the biological system is defined by the functional organization between the elements of matter that satisfies the optimum principle of the decrease in orgatropy.

More precisely, since the variation of the biological system in the course of development can be mathematically expressed by its *orgatropy*, the extremum principle for the O-FBS combines two variational properties, i.e. specialization and hierarchization, leading to a new state function, called the *functional order*, which indicates the *sense* of the variation of the system. *The biological system develops by decreasing its orgatropy, i.e. by increasing its functional order.* This brings out the profound difference between the physical universe and the biological world. In non-living matter, thermodynamic entropy cannot but increase with time, leading to an increase in molecular disorder. In contrast, the functional organization of a developing biological system is characterized by a decrease in orgatropy, corresponding to an increase in functional order. Thus, *the increase in the functional order of a biological system during its development, by means of the specialization and hierarchization of its structural units, is the second characteristic of life in non-living matter.*

The third condition of life in non-living matter is that the biological system must possess the geometrical properties essential for the operation of its functional dynamics, i.e. its D-FBS. This is a necessary condition because the product of the functional interaction between the structural units of the biological system is transported at a *finite velocity*. Since the activatory or inhibitory signal that is propagated from the source to the sink has to cover a certain distance in physical space in a given time, there is a significant time lag in the physiological action. The representation of this non-local action in a hierarchical system by a non-local field, satisfying the equation $H\psi = \Gamma$, includes the geometry of the biological system in the general formalism of the D-FBS. The propagation of such fields in matter has three properties that are specific to living organisms. Firstly, the propagation, defined by mathematical operators called *S-propagators*, describes the transport of biological signals through space in a hierarchical structure. Secondly, the propagation depends on the *connectivity* between the elements of matter, or the structural units, the connectivity being a mathematical function describing the relationship between the topology and the geometry of the biological system. Finally, the propagation defines the functional level of organization by means of its *time scale*, i.e. the dynamic behavior of the physiological function. *Thus, the propagation of non-local fields according to the connectivity of the structural units in a functional hierarchical system is the third characteristic of life in non-living matter.*

Before presenting a definition of life in terms of the functional organization of biological systems, let us briefly consider how the O-FBS and the D-FBS of a living organism vary with time. So far, we have been mainly concerned with the role of *biological space* in the living organism. The notion of the time-variation of living organisms naturally introduces the concept of *biological time*, which is evidently no less important and deserves at least as much attention as the role of space in - biology.

How biological systems vary with time

Living organisms go through two important stages that should be clearly distinguished — the developmental stage and the adult stage. During the development of a biological organism, there is evidently a profound relationship between the O-FBS and the D-FBS, which together constitute the biological system. This is reflected by the hypothesis of *self-association*, which postulates the existence of functional biological interactions on the grounds of the necessary stability of the dynamics of the D-FBS. Expressed as a *principle of vital coherence*, this corresponds to a *principle of conservation* of the functional dynamics of the biological system while it undergoes topological modification during its development. In other words, whenever new sources and sinks, i.e. new functional interactions, are created, the functional interactions of the O-FBS will be reorganized so as to increase the *domain of stability* of the dynamics of the physiological processes of the organism. Moreover, this reorganization, which implies the *specialization* and *hierarchization* of the O-FBS, will tend to increase the *functional order* of the biological system.

All this is, of course, a consequence of the expression of the genetic program of the living organism. Thus, the *density-connectivity* of the structural units of a biological system, which determines the propagation of *biological fields*, varies during its development. However, the O-FBS may also vary under other circumstances. For example, in the nervous system, the process of learning will obviously modify the system. It can be shown that a biological system, subjected to a condition of constant global organization, such as that of having a constant number of structural units, will undergo a redistribution of structural units between its different levels of structural organization. This will evidently affect the D-FBS and modify the propagation of biological fields. Furthermore, in the adult organism, there are also constraints of stability in the regulation of the physiological functions. We have seen that the *self-organization* of the biological system corresponds to the maintenance of the stability of the D-FBS when the O-FBS is modified. Thus, *the self-organization of a biological system is a consequence of the relationship between the topology and the dynamics of the system, i.e. between its structural and functional organizations.*

What happens when a perturbation of this relationship leads to modifications in the structure of a biological organism? In the case of structural modifications produced at the lowest molecular level of an organism, for example through errors of transcription of the genetic message, we have shown that the optimum regulation with respect to time between the different levels of structural organization can be determined by Hamiltonian dynamics (Chauvet, 1990). Structural modifications that occur during the functioning of the system may affect regulatory mechanisms, thus altering the functional organization, i.e. the topology of the system, and consequently, the associated biological fields. If the fluctuations produced have random effects at the lowest structural level, which also contains the largest number of structural units, then non-Hamiltonian dynamics, rather than Hamiltonian dynamics, will have to be used. In this case, the time-variation of the biological system is *irreversible*, as is clearly shown by the phenomenon of *aging* in a living organism. This is obviously a manifestation of *biological time*, i.e. the internal time of the individual organism on the time scale of its physiological processes.

One of the most fascinating aspects of life in non-living matter is, of course, the *evolution of the species*. The question therefore arises concerning the coherence of the principles of our theory of the integrated physiology of living organisms with those of evolution. In other words, is the time-variation of the organization of physiological functions compatible with intra-species and trans-species modifications? The time-variation of the phenomena of life at the level of an individual organism expresses the coupling of its topology and its dynamics on the time scale of its physiological processes. Similarly, the evolution of the species can be viewed as the consequence of a perturbation of the coupling of the topology and the dynamics of specific groups of biological organisms on the time scale of the living universe. Such a perturbation may be either an internal transformation compatible with the survival of the species, such that the physiological function is conserved, or else a reaction to the persistent influence of some particular property of the environment. In the former case, the biological system will undergo self-organization by means of successive adaptations without losing any of its physiological functions in the following generation (in keeping with the *restricted* principle of vital coherence), thus ensuring the conservation of the species. In the latter case, the evolution of the biological system involves the loss of some of its functions in the following generation through the modification of its genetic sex characteristics (according to the *general* principle of vital coherence), thus leading to trans-speciation. In both cases however, the physiological processes will be modified — from the point of view of the construction of the structural hierarchy — by the self-association of structural units during the development of living organisms. Then, if our theory is valid, the living universe, just like the individual organism, may be considered to operate on a specific time scale corresponding to its own aging process. In other words, the dynamics of the living universe and the individual organism operate on different time scales. All these ideas will be fully developed in a companion volume dealing specifically with the role of time in biology.

Finally, our theoretical approach to life in non-living matter leads to the following definition:

> *Life is produced by a hierarchical system of regulated processes representing the action of one hierarchical element of matter, or structural unit, on another. These processes, which are stabilized from a topological point of view, tend to increase the stability of the functional dynamics of a living organism by means of the self-association of structural units during development, and maintain this stability through time-optimum regulation between the hierarchical levels of functional organization.*

The coherence of this definition of life depends on the three fundamental axes that emerge from the observed behavior of biological systems. These are the functional organization (or the topology), the increase in functional stability produced by the self-association of elements of matter (or structural units) and the propagation of biological fields in hierarchically organized matter (or the dynamics of the physiological functions). In particular, each of these axes possesses a specific regulatory mechanism ensuring that *there is stabilization during development and stability during the functioning of a biological system.* Thus, whereas physical theories of matter are of a purely geometrical nature, biological theories must account for the geometrical as well as the topological nature of living organisms.

Some readers may have been discouraged by the degree of mathematical abstraction we have used in describing the fundamental phenomena of life. However, the ever-increasing precision of measuring instruments and the highly sophisticated techniques now available for physical and biological investigations are steadily pushing back the frontiers of the observable. The complexity of the universe as we understand it cannot but increase, justifying the construction of new integrative theories. Penrose (1989) has classified theories as being *superb, useful, tentative* or *misguided*. Among the rare theories considered superb, we have already referred to Maxwell's electromagnetic theory, Einstein's theories of relativity, and Planck's quantum theory. Useful theories include Weinberg and Salam's theory of weak interactions, and the Big Bang theory of the origin of the universe. Hawking's superstring theory and the Grand Unified Theory of astrophysics are among the tentative theories of modern science. However, Penrose gives no examples of any misguided theories, possibly out of charity. For our part, we can only hope that biologists will consider the ideas we have outlined here as a contribution to a tentative field theory of integrative physiology. Indeed, a new general physiology seems to be just around the corner, waiting to be discovered.

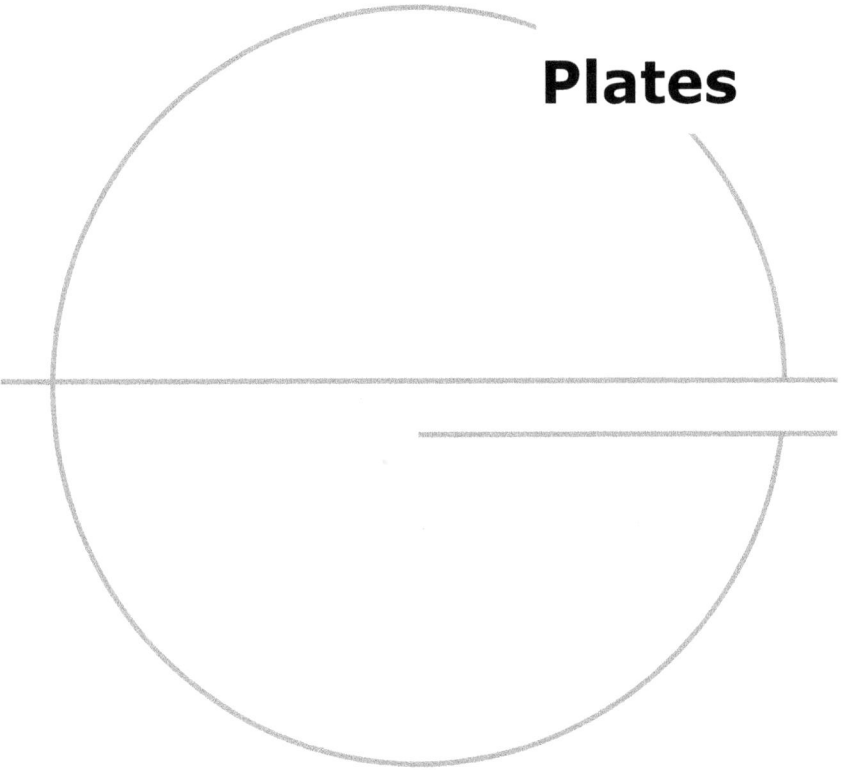

Plates

PLATES

Plate IV.1: Military architecture offers a good analogy of biological structural organization. The ramparts play the role of cellular and nuclear membranes. In this illustration, the constructions clearly date from very different periods but the functional hierarchy is carefully conserved.

Plate IV.2: This medieval representation of a besieged castle shows the functional organization of the interactions between the attackers and the defenders according to the structural organization shown in Plate IV.1. The military architects of the time considered the two lines of defense formed by the rampart and the fortification of the castle to be the best structure for defensive action.

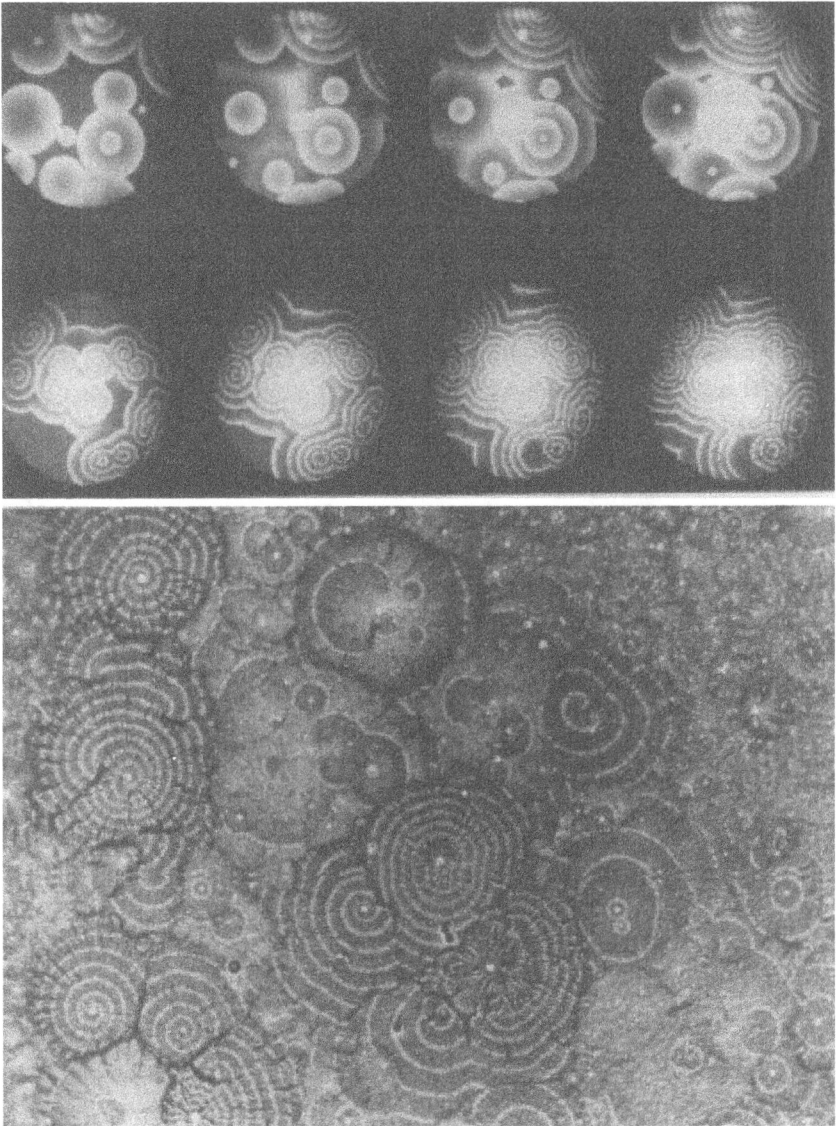

Plate V.1: Dissipative structures obtained with chemical reactions far from the state of equilibrium. These images have been interpreted by means of the thermodynamics of irreversible processes. The upper figure, showing circular waves at the top and spiral waves at the bottom (A. T. Winfree, 1980), illustrates the Belusov-Zhabotinsky reaction. The lower figure shows similar patterns found on *Dictyostelium discoideum* (P. C. Newell, 1983).

Plates

Plate V.2: Some of the extraordinary patterns observed in the evolution of living creatures. All these complex markings can be mathematically explained by means of reaction-diffusion equations (J. D. Murray, 1989).

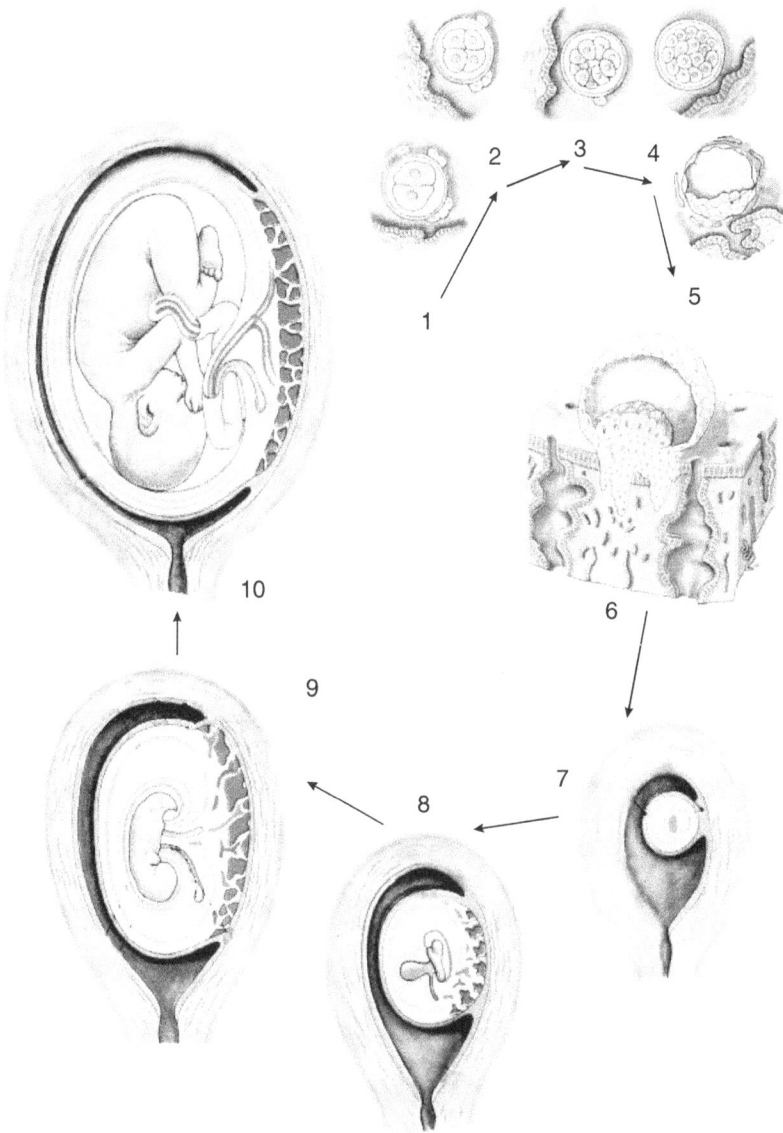

Plate V.3: Development of a human being from the stage of the first segmentation of the fecundated egg to that of the fully formed fetus. The stages
marked 1 to 4 are the same as those in Figure V.3.

Plates

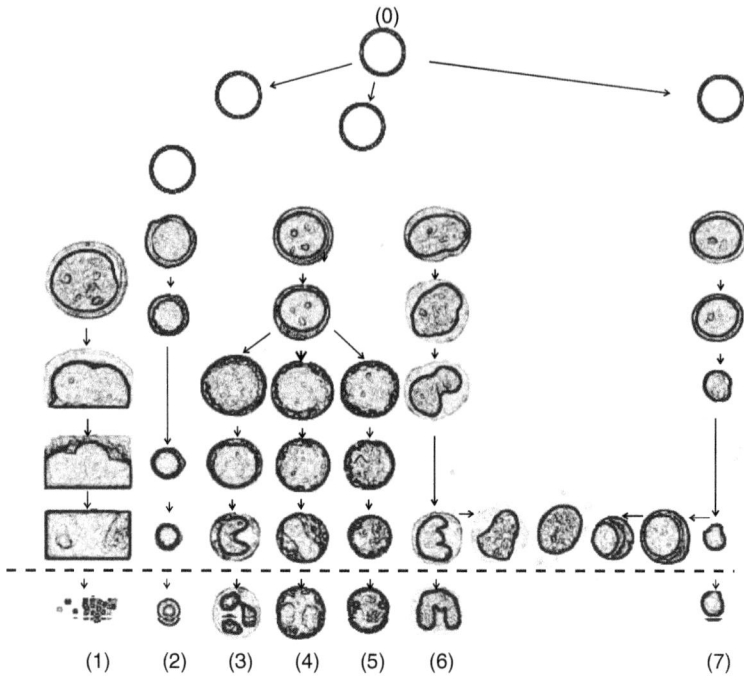

Plate V.4: This illustration of hematopoiesis shows the successive differentiation of blood cells from a single stem cell. The process leads to the formation of seven distinct types of cell: (1) blood platelets; (2) red blood cells; (3) neutrophils; (4) eosinophils; (5) basophils; (6) monocytes; and (7) lymphocytes.

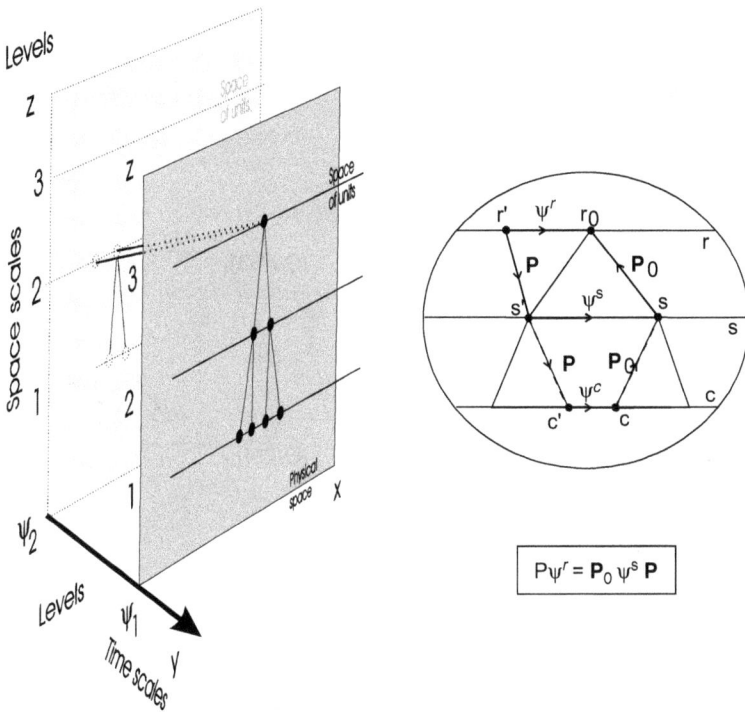

Plate VI.1: This figure illustrates the propagation of a functional interaction in the biological organism and its passage from one structural level to another. The inset represents the influence of synapses (s) and channels (c) on the propagation of ψ from r' to r_0 The functional interactions at the different levels are ψ^r, ψ^s and ψ^c. The S-propagators are noted $\mathbf{P}$ and $\mathbf{P}_0$.

In this example, we have considered two functions: the short-term propagation (measured in milliseconds) of the potential ψ, which involves the activity of short-term synapses, and μ, the synaptic efficacy (measured in seconds), which involves long-term synapses. The different time scales are shown on the y-axis, whereas the space scales that characterize the structure are shown on the x-axis. The roles played by time and space are thus visualized.

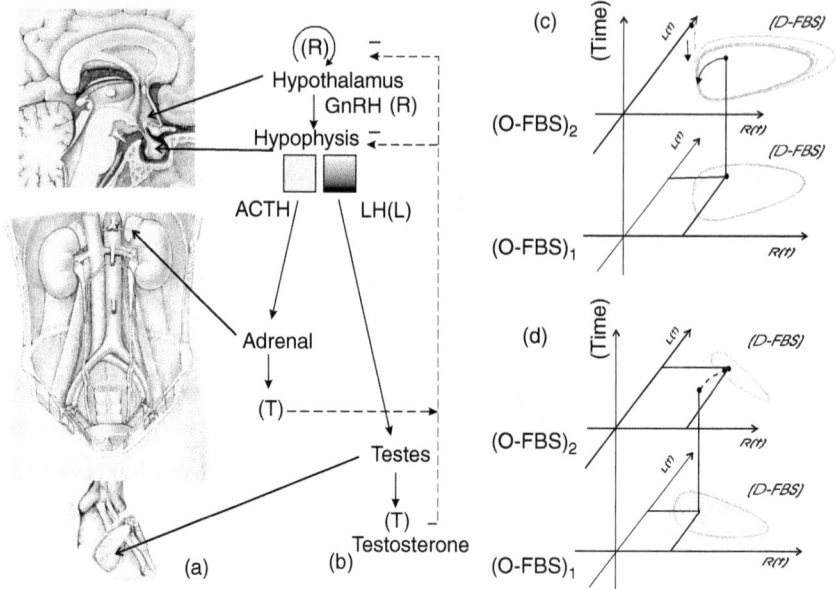

Plate VI.2: An illustration of self-organization provided by the regulation of the secretion of testosterone, the male sexual hormone. (a) The hypothalamo-hypophyso-testicular axis with its structural units; (b) the schema representing the endocrine regulation of testosterone; (c) the coupling of the topology and the dynamics of the normal phenomenon in the phase space (R, *L*), shown as limit cycle attractors; and (d) the modification of the limit cycle attractors following castration (see text for details).

REFERENCES

Atlan H (1972) *L'organisation Biologique et la Théorie de l'information*, Hermann, Paris.

Avery OT, MacLeod C and McCarty M (1944) Studies on the chemical nature of the substance inducing transformation of pneumococcal types, *Journal of Experimental Medicine* **79**, 137.

Babloyantz A and Nicolis G (1972) Chemical instabilities and multiple steady state transitions in Monod–Jacob type models, *Journal of Theoretical Biology* **34**, 185–192.

Baillet J and Nortier E (1992) *Précis de physiologie humaine*. Humaine, Ellipses, Paris.

Beadle G and Ephrussi B (1935) Différenciation de la couleur cinnabar chez la drosophile (*Drosophila melanogaster*), *Comptes Rendus de l'Academie des Sciences, Paris* **201**, 620.

Beadle G and Tatum EL (1941) Genetic control of biochemical reactions in *Neurospora*, *Proceedings of the National Academy of Sciences of the United States of America* **27**, 499.

Benzer S (1957) The elementary units of heredity, in McElroy and Glass (eds.) *The Chemical basis of Heredity*, John Hopkins Press, Baltimore, pp. 70.

Berge C (1983) *Graphes*, 3rd ed., Gauthier–Villars, Paris.

Bernard CI (1984) *Introduction à la Médecine Expérimentale*, Flammarion.

Beurle RL (1956) Properties of a mass of cells capable of regenerating pulses, *Philosophical Transactions of the Royal Society of London Series B-Biological Sciences* **240**, 55–94.

Brillouin L (1952) La vie, la pensée et la physico-chimie, in *Cahiers de la Pléiade*, *XIII* Gallimard, Paris.

Brown MC, Hopkins WG and Keynes RJ (1991) *Essentials of Neural Development*, Cambridge University Press, Cambridge.

Burger J, Machbub C and Chauvet G (1993) Numerical stability of interactive biological units, *Syst Anal - Model Simul* **13**, 197–207.

Cannon WB (1929) *Bodily changes in Pain, Hunger, Fear, and Rage,* 2nd ed., D. Appleton and Co., New York and London.

S Carnot (1824) *Réflexions sur la Puissance Motrice du Feu sur les Machines Propres a Developper cette Puissance*, Bachelier, Paris [Reflections on the motive power of heat and on machines fitted to develop this power, translated by R. H. Thurston, The American Society of Mechanical Engineers, New York (1943)].

Chauvet GA (1996) *Theoretical Systems in Biology: Hierarchical and Functional Integration*, Pergamon Press, Oxford.

References

Chauvet G (1974) *Méthode Préliminaire D'attribution des Fréquences Vibrationnelles. Application à Structures Dérivées de Composés Minéraux et Organiques.* PhD thesis, Nantes.

Chauvet G (1976) *Etude d'un Modèle Physique de L'écoulement non Permanent d'un Fluide dans les Voies Aériennes. Application aux Inégalités Ventilatoires,* MD thesis, Université d'Angers.

Chauvet G (1978a) Discussion on basic equations for a flow in human air passages and their representation, *International Journal of Bio-Medical Computing* **9**, 353–365.

Chauvet G (1978b) Mathematical study of flow in a non-symmetrical bronchial tree: application to asynchonous ventilation, *Mathematical Biosciences* **42**, 73–91.

Chauvet GA (1996a) *Theoretical Systems in Biology: Hierarchical and Functional Integration, Vol. I: Molecules and Cells,* Pergamon Press, Oxford.

Chauvet GA (1996b) *Theoretical Systems in Biology: Hierarchical and Functional Integration, Vol. II: Tissues and Organs,* Pergamon Press, Oxford.

Chauvet GA (1996c) *Theoretical Systems in Biology: Hierarchical and Functional Integration, Vol. III: Organisation and Regulation,* Pergamon Press, Oxford.

Chauvet GA (1993a) Hierarchical functional organization of formal biological systems: a dynamical approach. II. The concept of non-symmetry leads to a criterion of evolution deduced from an optimum principle of the (O-FBS) sub-system, *Philosophical Transactions of the Royal Society of London Series B-Biological Sciences* **339**, 445–461.

Chauvet GA (1993b) Non-locality in biological systems results from hierarchy: Application to the nervous system, *Journal of Mathematical Biology* **31**, 475–486.

Chauvet GA (1993c) Hierarchical functional organization of formal biological systems: a dynamical approach. III. The concept of non-locality leads to a field theory describing the dynamics at each level of organization of the (D-FBS) sub-system, *Philosophical Transactions of the Royal Society of London Series B-Biological Sciences* **339**, 463–481.

Chauvet GA (1993d) Hierarchical functional organization of formal biological systems: a dynamical approach. I. An increase of complexity by self-association increases the domain of stability of a biological system, *Philosophical Transactions of the Royal Society of London Series B-Biological Sciences* **339**, 425–444.

Chauvet GA (1993e) An n-Level Field Theory of Biological Neural Network, *Journal of Mathematical Biology* **31**, 771–795.

Chauvet GA and Costalat R (1995) On the functional organization in a biological structure: the example of enzyme organization, *C R Acad Sci Paris Série III* **318**, 529–535.

Chauvet P and Chauvet GA (1993) On the ability of cerebellar Purkinje units to constitute a neural network, in Erlbaum L (ed.) *WCNN'93: Proceedings of the World Congress on Neural Networks, Vol. 2, Portland, USA, July 11–15, 1993,* INNS Press, Portland pp. 602–605.

Chauvet GA (1999) S-propagators: A formalism for the hierarchical organization of physiological systems. Application to the nervous and the respiratory systems. *International Journal of General Systems* **28**(1), 53–96.

Chauvet GA (2002) On the mathematical integration of the nervous tissue based on the S-Propagator formalism. *Journal of Integrative Neuroscience* **1**(1), 31–68.

Coleman S (1985) *Aspects of symmetry,* Cambridge University Press, Cambridge.

D'Arcy T (1917) *On Growth and Form,* University Press, Cambridge.

References

Delattre P (1971a) *L'Evolution des Systèmes Moléculaires*, Maloine, Paris.

Delattre P (1971b) Topological and order properties in transformation systems, *Journal of Theoretical Biology* **32**, 269–282.

D'Espagnat B (1980) *A la Recherche du Réel. Le Regard d'un Physicienm*, Gauthier–Villars, Paris.

Einstein A (1989) *The Collected Papers of Albert Einstein, Vol. 2: The Swiss Years: Writings, 1900–1909*, Princeton, NJ.

Einstein A (1905) *On the Movement of Small Particles Suspended in Stationary Liquids Required by Molecular-Kinetic Theory of Heat.*

Einstein A (1915) *The Field Equations for Gravitation*, Sitzungsberichte der Deutschen Akademie der Wissenschaften zu Berlin, Klasse für Mathematik, Physik und Technik, pp. 844.

Einstein A (1923) *The Principle of Relativity*, Methuen & Co, London.

Einstein A, Lorentz HA, Weyl H and Minkowski H (1952) *The Principle of Relativity*,Dover Publications, Inc., New York.

Feynman R (1969) *The Feyman Lectures on Physics*, Addison–Wesley, Reading, MA.

Flemming W (1882) *Zellsubstansz, Kern und Zelltheilung*, Leipzig.

Geymonat L (1992) Galilée, *Editions du Seuil*, Paris.

Glansdorff P and Prigogine I (1971) *Thermodynamic Theory of Structure Stability and Fluctuations*, Wiley and Sons, London.

Goldbeter A (1974) Modulation of the adenylate energy charge by sustained metabolic oscillations, *FEBS Letters* **43**, 327–330.

Goldbeter A (1977) On the role of enzyme cooperativity in metabolic oscillation: analysis of the Hill coefficient in a model for glycolytic periodicities, *Biophysical Chemistry* **6**, 95–99.

Goldbeter A and Lefever R (1972) Dissipative structures for an allosteric model. Application to glycolytic oscillations, *Biophysical Journal* **12**, 1302–1315.

Goodwin BC (1978) *Analytical Cell Physiology*, Academic Press, New York.

Griffith F (1922) Types of pneumococci obtained from cases of lobar pneumonia, *Reports on Public Health and Medical Subjects, Bacteriological studies, HMSO, London*, **13**.

Griffith JS (1963) A field theory of neural nets. I: Derivation of field equations, *Bulletin of Mathematical Biophysics* **25**, 111–120.

Grodins FS, Buell J and Bart AJ (1967) Mathematical analysis and digital simulation of the respiratory control system, *Journal of Applied Physiology* **22**, 260–276.

Gros F (1986) *Les Secrets du Gène*, Odile Jacob, Paris.

Guyton AC, Coleman TG and Granger HJ (1972) Circulation: overall regulation, *Annual Review of Physiology* **34**, 13–46.

Hawking SW (1988) *A Brief History of Time. From the Big Bang to Black Holes*, Bantam Book, New York.

Heinrich R and Rapoport TA (1976) The regulatory principles of the glycolysis of erythrocytes *in vivo* and *in vitro*, in Keleti T and Lakatos S (eds.) *Mathematical Models of Metabolic Regulation*, Adademiai Kiado, Budapest.

Hess B and Boiteux A (1971) Oscillatory phenomena in biochemistry, *Annual Review of Biochemistry* **40**, 237–258.

Hodgkin AL and Huxley AF (1952) A quantitative description of membrane current and its application to conduction and excitation in nerve, *Journal of Physiology* **117**, 500–544.

References

Hopfield JJ (1982) Neural networks and physical systems with emergent collective computational abilities, *Proceedings of the National Academy of Sciences of the United States of America* **79**, 2554–2558.

Jacob F and Monod J (1961) Genetic regulatory mechanisms in the synthesis of proteins, *Journal of Molecular Biology* **3**, 318–356.

Jacquez JA (1979) *Respiratory Physiology*, McGraw Hill, New York.

Jacquez JA (1985) *Compartmental Analysis in Biology and Medicine*, 2nd ed., University of Michigan Press, Ann Arbor.

Koshland DE Jr, Némethy G and Filmer D (1966) Comparison of experimental binding data and theoretical models in proteins containing subunits, *Biochemistry* **5**, 365–385.

Koushanpour E and Stipp GK (1982) Mathematical simulation of the body fluid regulating system in dog, *Journal of Theoretical Biology* **99**, 203–235.

Leech JW (1965) *Classical Mechanics*, Methuen and Co.

Leyden (1638) is the city where the book of Galilee was published.

Maxwell JC (1872) *Treatise on Electricity and Magnetism*, London.

McClintock B (1995) Controlled mutation in maize, Carnegie Inst. Wash., Tear Book, pp. 54, 245.

Mendel G (1866) Versuchen über Pflanzen-Hybriden, *Verhandlungen der Naturforschenden* **4**, 3.

Meselson M and Stahl F (1958) The replication of DNA in *Escherichia coli*, *Proceedings of the National Academy of Sciences of the United States of America* **44**, 671.

Monod J, Wyman J and Changeux J-P (1965) On the nature of allosteric transitions: a plausible model, *Journal of Molecular Biology* **12**, 88–118.

Morgan TH (1910) Sex limited inheritance in *Drosophila*, *Science* **32**, 120.

Murray JD (1989) *Mathematical Biology*, Heidelberg, Springer–Verlag.

Newell PC (1983) Attraction and adhesion in the slime mold *Dictyostelium*, in Smith JE (ed.) *Fungal Differentiation: A Contemporary Synthesis*, Mycology series 43, Marcel Dekker, New York, pp. 43–71.

Newton I (1687) Philosophiae naturalis principia mathematica, *The Principia*.

Newton I (1686) *Principia. Vol. I: The Motion of Bodies*, 3rd ed., University of California Press, Berkeley.

Oster GF, Perelson AS and Katchalsky A (1973) Network thermodynamics: dynamic modelling of biological systems, *Quarterly Review of Biophysics* **6**, 1–134.

Ovádi J (1991) Physiological significance of metabolic channeling, *Journal of Theoretical Biology* **152**, 1–22.

Paiva M (1973) Gaz transport in the human lung, *Journal of Applied Physiology* **35**, 401–410.

Penrose R (1989) *The Emperor's New Mind*, Oxford University Press, New York.

Poincaré H (1890) Sur le problème des trois corps et les équations de la dynamique, *Acta Mathematica* **13**, 1–270.

Poincaré H (1968) *La Science et l'Hypothèse*, Flammarion, Paris.

Poincaré H (1980) *Oeuvres, 12 volumes*, Gauthier–Villars, Paris.

Prigogine I, Nicolis G and Babloyantz A (1972) Thermodynamics of evolution. The ideas of nonequilibrium order and of the search for stability extend Darwin's concept back to the prebiotic stage by redefining the "fittest", *Physics Today* **December**, 38–44.

References

Queiroz O (1979) Modélisation du fonctionnement cellulaire dans un contexte physiologique: Quand? Pourquoi? Pour quoi faire?, in Delattre P and Thellier M (eds.), *Elaboration et Justification des Modèles*, Maloine, Paris.

Rashevsky N (1961) *Mathematical Principles in Biology and their Applications*, Charles C. Thomas, Springfield, ILL.

Roy B (1970) *Algèbre Moderne et Théorie des Graphes*, Dunod, Paris.

Savageau MA (1974) Comparison of the classical and autogenous system of regulation in inducible operons, *Nature* **252**, 546–549.

Savageau MA (1979) Allometric morphogenesis of complex systems: derivation of the basic equations from first principles, *Proceedings of the National Academy of Sciences of the United States of America* **76**, 6023–6025.

Schrodinger E (1945) *What is Life?* Cambridge University Press, Cambridge.

Shepherd GM (1992) Canonical Neurons and their Computational Organization, in McKenna T, Davis J and Zornetzer SF (eds.), *Single Neuron Computation*, Academic Press Inc., pp. 27–60.

Silbernagl S and Despopoulos A (2003) *Color Atlas of Physiology*, Theime Medical Publishers.

Stryer L (1992) *Biochemistry*, 3rd ed., WH Freeman and Co., New York.

Thom R (1972) *Stabilité Structurelle et Morphogenèse*, Ediscience, Paris.

Ullmo J (1969) *La Pensée Scientifique Moderne*, Flammarion, Paris.

Waldeyer W (1902) Die protozoen und die zelletheorie, *Archiv für Protistenkunde* **1**, 1.

Watson J (2001) *Molecular Biology of the Gene,* Benjamin-Cummings Publishing Company.

Watson JD and Crick FH (1953) Molecular structure of nucleic acids. A structure for desoxyribose nucleic acid, *Nature* **171**, 737.

Weibel ER (1963) *Morphometry of the Human Lung*, Academic Press, New York.

Westerhoff HV and Welch GR (1992) Enzyme organization and the direction of metabolic flow: physico-chemical considerations, *Current Topics in Cellular Regulation* **33**, 361–390.

Wigner EP (1970) *Symmetries and reflections, Scientific Essays*, 2nd ed., Cambridge, MIT Press, Massachusetts.

Wilson EB, Decius JC and Cross PC (1955) *Molecular Vibrations, The Theory of Infrared and Raman Vibrational Spectra*, Dover, New York.

Winfree AT (1980) *The Geometry of Biological Time*, Springer–Verlag, Berlin.

Wyman J Jr. (1964) Linked functions and reciprocal effects in hemoglobin: a second look, *Advances in Protein Chemistry* **19**, 223–286.

Yates FE (1982) Book review: Physiological Integrations in Action, *American Journal of Physiology* **12**, R471–R473.

INDEX

Index